Numerical Methods

Numerical Methods

Editors

Lorentz Jäntschi
Daniela Roșca

MDPI • Basel • Beijing • Wuhan • Barcelona • Belgrade • Manchester • Tokyo • Cluj • Tianjin

Editors
Lorentz Jäntschi
Technical University of Cluj-Napoca
Romania

Daniela Roșca
Technical University of Cluj-Napoca
Romania

Editorial Office
MDPI
St. Alban-Anlage 66 4052 Basel,
Switzerland

This is a reprint of articles from the Special Issue published online in the open access journal *Mathematics* (ISSN 2227-7390) (available at: https://www.mdpi.com/journal/mathematics/special_issues/Numerical_Methods_2020).

For citation purposes, cite each article independently as indicated on the article page online and as indicated below:

LastName, A.A.; LastName, B.B.; LastName, C.C. Article Title. *Journal Name* **Year**, *Article Number*, Page Range.

ISBN 978-3-03943-318-6 (Hbk)
ISBN 978-3-03943-319-3 (PDF)

© 2020 by the authors. Articles in this book are Open Access and distributed under the Creative Commons Attribution (CC BY) license, which allows users to download, copy and build upon published articles, as long as the author and publisher are properly credited, which ensures maximum dissemination and a wider impact of our publications.

The book as a whole is distributed by MDPI under the terms and conditions of the Creative Commons license CC BY-NC-ND.

Contents

About the Editors vii

Preface to "Numerical Methods" ix

Lorentz Jäntschi
Detecting Extreme Values with Order Statistics in Samples from Continuous Distributions
Reprinted from: *Mathematics* **2020**, *8*, 216, doi:10.3390/math8020216 1

Monica Dessole, Fabio Marcuzzi and Marco Vianello
dCATCH—A Numerical Package for d-Variate near G-Optimal Tchakaloff Regression via Fast NNLS
Reprinted from: *Mathematics* **2020**, *8*, 1122, doi:10.3390/math8071122 23

Soledad Moreno-Pulido, Francisco Javier Garcia-Pacheco, Clemente Cobos-Sanchez and Alberto Sanchez-Alzola
Exact Solutions to the Maxmin Problem $\max \|Ax\|$ Subject to $\|Bx\| \leq 1$
Reprinted from: *Mathematics* **2020**, *8*, 85, doi:10.3390/math8010085 39

Kin Keung Lai, Shashi Kant Mishra and Bhagwat Ram
On q-Quasi-Newton's Method for Unconstrained Multiobjective Optimization Problems
Reprinted from: *Mathematics* **2020**, *8*, 616, doi:10.3390/math8040616 65

Deepak Kumar, Janak Raj Sharma and Lorentz Jäntschi
Convergence Analysis and Complex Geometry of an Efficient Derivative-Free Iterative Method
Reprinted from: *Mathematics* **2019**, *7*, 919, doi:10.3390/math7100919 79

Janak Raj Sharma, Sunil Kumar and Lorentz Jäntschi
On Derivative Free Multiple-Root Finders with Optimal Fourth Order Convergence
Reprinted from: *Mathematics* **2020**, *8*, 1091, doi:10.3390/math8071091 91

Ampol Duangpan, Ratinan Boonklurb and Tawikan Treeyaprasert
Finite Integration Method with Shifted Chebyshev Polynomials for Solving Time-Fractional Burgers' Equations
Reprinted from: *Mathematics* **2019**, *7*, 1201, doi:10.3390/math7121201 107

Adrian Holhoş and Daniela Roşca
Orhonormal Wavelet Bases on The 3D Ball Via Volume Preserving Map from The Regular Octahedron
Reprinted from: *Mathematics* **2020**, *8*, 994, doi:10.3390/math8060994 131

Jintae Park, Sungha Yoon, Chaeyoung Lee and Junseok Kim
A Simple Method for Network Visualization
Reprinted from: *Mathematics* **2020**, *8*, 1020, doi:10.3390/math8061020 147

SAIRA, Shuhuang Xiang, Guidong Liu
Numerical Solution of the Cauchy-Type Singular Integral Equation with a Highly Oscillatory Kernel Function
Reprinted from: *Mathematics* **2019**, *7*, 872, doi:10.3390/math7100872 161

About the Editors

Lorentz Jäntschi was born in Făgăraș, Romania, in 1973. In 1991, he moved to Cluj-Napoca, Cluj, where he completed his studies. In 1995, he was awarded his B.Sc. and M.Sc. in Informatics (under the supervision of Prof. Militon FRENȚIU); in 1997, his B.Sc. and M.Sc. in Physics and Chemistry (under the supervision of Prof. Theodor HODIȘAN); in 2000, his Ph.D. in Chemistry (under the supervision of Prof. Mircea V. DIUDEA); in 2002, his M.Sc. in Agriculture (under the supervision of Prof. Iustin GHIZDAVU and Prof. Mircea V. DIUDEA); and in 2010, his Ph.D. in Horticulture (under the supervision of Prof. Radu E. SESTRAȘ). In 2013, he conducted a postdoc in Horticulture (with Prof. Radu E. SESTRAȘ) and that same year (2013), he became a Full Professor of Chemistry at the Technical University of Cluj-Napoca and Associate at Babes-Bolyai University, where he advises on Ph.D. studies in Chemistry. He currently holds both of these positions. Throughout his career, he has conducted his research and education activities under the auspices of various institutions: the G. Barițiu (1995–1999) and Bălcescu (1999–2001) National Colleges, the Iuliu Hațieganu University of Medicine and Pharmacy (2007–2012), Oradea University (2013–2015), and the Institute of Agricultural Sciences and Veterinary Medicine at University of Cluj-Napoca (2011–2016). He serves as Editor for the journals Notulae Scientia Biologicae, Notulae Horti Agro Botanici Cluj-Napoca, Open Agriculture, and Symmetry. He has served as Editor-in-Chief of the Leonardo Journal of Sciences and the Leonardo Electronic Journal of Practices and Technologies (2002–2018) and as Guest Editor (2019–2020) of Mathematics.

Daniela Roșca was born in Cluj-Napoca, Romania in 1972. In 1995, she was awarded her B.Sc. in Mathematics, and in 1996, her M.Sc. in Mathematics (Numerical and Statistical Calculus). In 2004, she became Doctor in Mathematics with a thesis entitled "Approximation with Wavelets" (defended: January 9th, 2004) and conducted a postdoc in Computing in 2013 (with Prof. Sergiu NEDEVSCHI). That same year (2013), she became a Full Professor of Mathematics at the Technical University of Cluj-Napoca, where she advises on Ph.D. studies in Mathematics. She was Invited Professor at Université Catholique de Louvain, Louvain-la-Neuve, Belgium on numerous occasions (13–27 January 2011 and 10–24 January 2013 and twice for 2 weeks in the academic years 2006–2007, 2007–2008, 2008–2009, and 2009–2010) delivering courses and seminars for the 3rd cycle (doctoral school) on wavelet analysis on the sphere and other manifolds.

Preface to "Numerical Methods"

The Special Issue "Numerical Methods" (2020) was open for submissions in 2019–2020) and welcomed papers from broad interdisciplinary areas since 'numerical methods' are a specific form of mathematics that involve creating and using algorithms to map out the mathematical core of a practical problem. Numerical methods naturally find application in all fields of engineering, physical sciences, life sciences, social sciences, medicine, business, and even arts. The common uses of numerical methods include approximation, simulation, and estimation, and there is almost no scientific field in which numerical methods do not find a use.

Some subjects included in 'numerical methods' are IEEE arithmetic, root finding, systems of equations, least squares estimation, maximum likelihood estimation, interpolation, numeric integration, and differentiation—the list may go on and on. Mathematical subject classification for numerical methods includes topics in conformal mapping theory in connection with discrete potential theory and computational methods for stochastic equations, but most of the subjects are within approximation methods and numerical treatment of dynamical systems, numerical methods, and numerical analysis. Also included are topics in numerical methods for deformable solids, basic methods in fluid mechanics, basic methods for optics and electromagnetic theory, basic methods for classical thermodynamics and heat transfer, equilibrium statistical mechanics, time-dependent statistical mechanics, and last but not least, mathematical finance. In short, the topics of interest deal mainly with numerical methods for approximation, simulation, and estimation. The deadline for manuscript submissions was closed on 30 June 2020.

Considering the importance of numerical methods, two representative examples should be given. First, the Jenkins–Traub method (published as "Algorithm 419: Zeros of a Complex Polynomial" and "Algorithm 493: Zeros of a Real Polynomial") which practically put the use of computers to another level in numerical problems. Second, the Monte Carlo method (published as "'he Monte-Carlo Method") which gave birth to the broad class of computational algorithms found today that rely on repeated random sampling to obtain numerical results. Today, the "numerical method" topic is much more diversified than 50 years ago, especially because of the technological progress and this series of collected papers is proof of this fact.

Results communicated here include topics ranging from statistics (Detecting Extreme Values with Order Statistics in Samples from Continuous Distributions, https://www.mdpi.com/2227-7390/8/2/216) and statistical software packages (dCATCH—A Numerical Package for d-Variate near G-Optimal Tchakaloff Regression via Fast NNLS, https://www.mdpi.com/2227-7390/8/7/1122) to new approaches for numerical solutions (Exact Solutions to the Maxmin Problem max $\|Ax\|$ Subject to $\|Bx\| \leq 1$, https://www.mdpi.com/2227-7390/8/1/85; On q-Quasi-Newton's Method for Unconstrained Multiobjective Optimization Problems, https://www.mdpi.com/2227-7390/8/4/616; Convergence Analysis and Complex Geometry of an Efficient Derivative-Free Iterative Method, https://www.mdpi.com/2227-7390/7/10/919; On Derivative Free Multiple-Root Finders with Optimal Fourth Order Convergence, https://www.mdpi.com/2227-7390/8/7/1091; Finite Integration Method with Shifted Chebyshev Polynomials for Solving Time-Fractional Burgers' Equations, https://www.mdpi.com/2227-7390/7/12/1201) to the use of wavelets (Orhonormal Wavelet Bases on The 3D Ball Via Volume Preserving Map from the Regular Octahedron, https://www.mdpi.com/2227-7390/8/6/994) and methods for visualization (A Simple Method for

Network Visualization, https://www.mdpi.com/2227-7390/8/6/1020).

Lorentz Jäntschi, Daniela Roşca
Editors

Article
Detecting Extreme Values with Order Statistics in Samples from Continuous Distributions

Lorentz Jäntschi [1,2]

[1] Department of Physics and Chemistry, Technical University of Cluj-Napoca, Cluj-Napoca 400641, Romania; lorentz.jantschi@chem.utcluj.ro or lorentz.jantschi@ubbcluj.ro
[2] Institute of Doctoral Studies, Babeş-Bolyai University, Cluj-Napoca 400091, Romania

Received: 17 December 2019; Accepted: 4 February 2020; Published: 8 February 2020

Abstract: In the subject of statistics for engineering, physics, computer science, chemistry, and earth sciences, one of the sampling challenges is the accuracy, or, in other words, how representative the sample is of the population from which it was drawn. A series of statistics were developed to measure the departure between the population (theoretical) and the sample (observed) distributions. Another connected issue is the presence of extreme values—possible observations that may have been wrongly collected—which do not belong to the population selected for study. By subjecting those two issues to study, we hereby propose a new statistic for assessing the quality of sampling intended to be used for any continuous distribution. Depending on the sample size, the proposed statistic is operational for known distributions (with a known probability density function) and provides the risk of being in error while assuming that a certain sample has been drawn from a population. A strategy for sample analysis, by analyzing the information about quality of the sampling provided by the order statistics in use, is proposed. A case study was conducted assessing the quality of sampling for ten cases, the latter being used to provide a pattern analysis of the statistics.

Keywords: probability computing; Monte Carlo simulation; order statistics; extreme values; outliers

MSC: 62G30; 62G32; 62H10; 65C60

1. Introduction

Under the assumption that a sample of size n, was drawn from a certain population ($x_1, ..., x_n \in X$) with a known distribution (with known probability density function, PDF) but with unknown parameters (in number of m, $\{\pi_1, ..., \pi_m\}$), there are alternatives available in order to assess the quality of sampling.

One category of alternatives sees the sample as a whole—and in this case, a series of statistics was developed to measure the agreement between a theoretical (in the population) and observed (of the sample) distribution. This approach is actually a reversed engineering of the sampling distribution, providing a likelihood for observing the sample as drawn from the population. To do this for any continuous distribution, the problem is translated into the probability space by the use of a cumulative distribution function (CDF).

Formally, if $\text{PDF}(x; (\pi_j)_{1 \leq j \leq m})$ takes values on a domain D, then CDF is defined by Equation (1) and $\{p_1, ..., p_n\}$ defined by Equation (2) is the series of cumulative probabilities associated with the drawings from the sample.

$$\text{CDF}(x; (\pi_j)_{1\leq j\leq m}) = \int_{\inf(D)}^{x} \text{PDF}(t; (\pi_j)_{1\leq j\leq m}) dt \tag{1}$$

$$\{p_1, ..., p_n\} = \text{CDF}(\{x_1, ..., x_n\}; (\pi_j)_{1\leq j\leq m}). \tag{2}$$

CDF is always a bijective (and invertible; let InvCDF be its inverse, Equation (3)) function.

$$x = \text{InvCDF}(p; (\pi_j)_{1 \leq j \leq m}). \quad (3)$$

The series of cumulative probabilities $\{p_1, ..., p_n\}$, independently of the distribution (PDF) of the population (X) subjected to the analysis, have a known domain ($0 \leq p_i \leq 1$ for all $1 \leq i \leq n$) belonging to the continuous uniform distribution ($p_1, ..., p_n \in U(0,1)$). In the sorted cumulative probabilities ($\{q_1, ..., q_n\}$ defined by Equation (4)), sorting defines an order relationship ($0 \leq q_1 \leq ... \leq q_n \leq 1$).

$$\{q_1, ..., q_n\} = \text{SORT}(\{p_1, ..., p_n\}; \text{"ascending"}). \quad (4)$$

If the order of drawing in sample ($\{x_1, ..., x_n\}$) and of appearance in the series of associated CDF ($\{p_1, ..., p_n\}$) is not relevant (e.g., the elements in those sets are indistinguishable), the order relationship defined by Equation (4) makes them ($\{q_1, ..., q_n\}$) distinguishable (the order being relevant).

A series of order statistics (OS) were developed (to operate on ordered cumulative probabilities $\{q_1, ..., q_n\}$) and they may be used to assess the quality of sampling for the sample taken as a whole (Equations (5)–(10) below): Cramér–von Mises ($CM_{Statistic}$ in Equation (5), see [1,2]), Watson U2 ($WU_{Statistic}$ in Equation (6), see [3]), Kolmogorov–Smirnov ($KS_{Statistic}$ in Equation (7), see [4–6]), Kuiper V ($KV_{Statistic}$ in Equation (8), see [7]), Anderson–Darling ($AD_{Statistic}$ in Equation (9), see [8,9]), and H1 ($H1_{Statistic}$ in Equation (10), see [10]).

$$CM_{Statistic} = \frac{1}{12n} + \sum_{i=1}^{n} \left(\frac{2i-1}{2n} - q_i\right)^2 \quad (5)$$

$$WU_{Statistic} = CM_{Statistic} + \left(\frac{1}{2} - \frac{1}{n}\sum_{i=1}^{n} q_i\right)^2 \quad (6)$$

$$KS_{Statistic} = \sqrt{n} \cdot \max_{1 \leq i \leq n}\left(q_i - \frac{i-1}{n}, \frac{i}{n} - q_i\right) \quad (7)$$

$$KV_{Statistic} = \sqrt{n} \cdot \left(\max_{1 \leq i \leq n}\left(q_i - \frac{i-1}{n}\right) + \max_{1 \leq i \leq n}\left(\frac{i}{n} - q_i\right)\right) \quad (8)$$

$$AD_{Statistic} = -n - \frac{1}{n}\sum_{i=1}^{n}(2i-1)\ln\left(q_i(1-q_{n-i})\right) \quad (9)$$

$$H1_{Statistic} = -\sum_{i=1}^{n} q_i \ln(q_i) - \sum_{i=1}^{n}(1-q_i)\ln(1-q_i). \quad (10)$$

Recent uses of those statistics include [11] (CM), [12] (WU), [13] (KS), [14] (AD), and [15] (H1). Any of the above given test statistics are to be used, providing a risk of being in error for the assumption (or a likelihood to observe) that the sample ($\{x_1, ..., x_n\}$) was drawn from the population (X). Usually this risk of being in error is obtained from Monte Carlo simulations (see [16]) applied on the statistic in question and, in some of the fortunate cases, there is also a closed-form expression (or at least, an analytic expression) for CDF of the statistic available as well. In the less fortunate cases, only 'critical values' (values of the statistic for certain risks of being in error) for the statistic are available.

The other alternative in assessing the quality of sampling refers to an individual observation in the sample, specifically the less likely one (having associated q_1 or q_n with the notations given in Equation (4)). The test statistic is g1 [15], given in Equation (11).

$$g1_{Statistic} = \max_{1 \leq i \leq n} |p_i - 0.5|. \quad (11)$$

It should be noted that 'taken as a whole' refers to the way in which the information contained in the sample is processed in order to provide the outcome. In this scenario ('as a whole'), the entirety of the information contained in the sample is used. As it can be observed in Equations (5)–(10), each formula uses all values of sorted probabilities ($\{q_1, ..., q_n\}$) associated with the values ($\{x_1, ..., x_n\}$) contained in the sample, while, as it can be observed in Equation (11), only the extreme value ($\max(\{q_1, ..., q_n\})$ or $\min(\{q_1, ..., q_n\})$) is used; therefore, one may say that only an individual observation (the extremum portion of the sample) yields the statistical outcome.

The statistic defined by Equation (11) no longer requires cumulative probabilities to be sorted; one only needs to find the most departed probability from 0.5—see Equation (11)—or, alternatively, to find the smallest (one having associated q_1 defined by Equation (4)) and the largest (one having associated q_n defined by Equation (4)), and to find which deviates from 0.5 the most ($g1_{Statistic} = \max\{|q_1 - 0.5|, |q_n - 0.5|\}$).

We hereby propose a hybrid alternative, a test statistic (let us call it TS) intended to be used in assessing the quality of sampling for the sample, which is mainly based on the less likely observation in the sample, Equation (12).

$$TS_{Statistic} = \frac{\max_{1 \leq i \leq n} |p_i - 0.5|}{\sum_{1 \leq i \leq n} |p_i - 0.5|}. \tag{12}$$

The aim of this paper is to characterize the newly proposed test statistic (TS) and to analyze its peculiarities. Unlike the test statistics assessing the quality of sampling for the sample taken as a whole (Equations (5)–(10)), and like the test statistic assessing the quality of sampling based on the less likely observation of the sample, Equation (11), the proposed statistic, Equation (12), does not require that the values or their associated probabilities ($\{p_1, ..., p_n\}$) be sorted (as $\{q_1, ..., q_n\}$); since (like the g1 statistic) it uses the extreme value from the sample, one can still consider it a sort of OS [17]. When dealing with extreme values, the newly proposed statistic, Equation (12), is a much more natural construction of a statistic than the ones previously reported in the literature, Equations (5)–(10), since its value is fed mainly from the extreme value in the sample (see the *max* function in Equation (12)). Later, it will be given a pattern analysis, revealing that it belongs to a distinct group of statistics that are more sensitive to the presence of extreme values. A strategy of using the pool of OS (Equations (5)–(12)) including TS in the context of dealing with extreme values is given, and the probability patterns provided by the statistics are analyzed.

The rest of the paper is organized as follows. The general strategy of sampling a CDF from an OS and the method of combining probabilities from independent tests are given in Section 2, while the analytical formula for the proposed statistic is given in Section 3.1, and computation issues and proof of fact results are given in Section 3.2. Its approximation with other functions is given in Section 3.3. Combining its calculated risk of being in error with the risks from other statistics is given in Section 3.4, while discussion of the results is continued with a cluster analysis in Section 3.5, and in connection with other approaches in Section 3.6. The paper also includes an appendix of the source codes for two programs and accompanying Supplementary Material.

2. Material and Method

2.1. Addressing the Computation of CDF for OS(s)

A method of constructing the observed distribution of the g1 statistic, Equation (11), has already been reported elsewhere [15]. A method of constructing the observed distribution of the Anderson–Darling (AD) statistic, Equation (9), has already been reported elsewhere [17]; the method for constructing the observed distribution of any OS via Monte Carlo (MC) simulation, Equations (5)–(12), is described here and it is used for TS, Equation (12).

Let us take a sample size of n. The MC simulation needs to generate a large number of samples (let the number of samples be m) drawn from uniform continuous distribution ($\{p_1, ..., p_n\}$ in

Equation (2)). To ensure a good quality MC simulation, simply using a random number generator is not good enough. The next step (Equations (10)–(12) do not require this) is to sort the probabilities to arrive at $\{q_1, ..., q_n\}$ from Equation (4) and to calculate an OS (an order statistic) associated with each sample. Finally, this series of sample statistics ($\{OS_1, ..., OS_w\}$ in Figure 1) must be sorted in order to arrive at the population emulated distribution. Then, a series of evenly spaced points (from 0 to 1000 in Figure 1) corresponding to fixed probabilities (from $InvCDF_0 = 0$ to $InvCDF_{1000} = 1$ in Figure 1) is to be used saving the (OS statistic, its observed CDF probability) pairs (Figure 1).

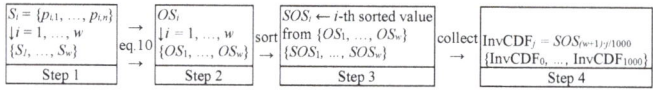

Figure 1. The four steps to arrive at the observed CDF of OS.

The main idea is how to generate a good pool of random samples from a uniform $U(0,1)$ distribution. Imagine a (pseudo) random number generator, *Rand*, is available, which generates numbers from a uniform $U(0,1)$ distribution, from a $[0,1)$ interval; such an engine is available in many types of software and in most cases, it is based on Mersenne Twister [18]. What if we have to extract a sample of size $n = 2$? If we split in two the $[0,1)$ interval (then into $[0, 0.5)$ and $[0.5, 1)$) then for two values (let us say $v1$ and $v2$), the contingency of the cases is illustrated in Figure 2.

$\in [0, 1)$	v_1		v_2		$v_1 v_2$				$v_1 + v_2$		
$\in [0, 0.5)$	0	1	0	1	00	01	10	11	0	1	2
occurrence	50%	50%	50%	50%	25%	25%	25%	25%	25%	50%	25%

Figure 2. Contingency of two consecutive drawings from $[0, 1)$.

According to the design given in Figure 2, for 4 (=2²) drawings of two numbers ($v1$ and $v2$) from the $[0, 1)$ interval, a better uniform extraction ($v1v2$, 'distinguishable') is ("00") to extract first ($v1$) from $[0, 0.5)$ and second ($v2$) from $[0, 0.5)$, then ("01") to extract first ($v1$) from $[0, 0.5)$ and second ($v2$) from $[0.5, 1)$, then ("10") to extract first ($v1$) from $[0, 0.5)$ and second ($v2$) from $[0.5, 1)$, and finally ("11") to extract first ($v1$) from $[0.5, 1)$ and second ($v2$) from $[0.5, 1)$.

An even better alternative is to do only 3 (=2 + 1) drawings ($v1 + v2$, 'undistinguishable'), which is ("0") to extract both from $[0, 0.5)$, then "1" to extract one (let us say first) from $[0, 0.5)$, and another (let us say second) from $[0.5, 1)$, and finally, ("2") to extract both from $[0.5, 1)$ and to keep a record for their occurrences (1, 2, 1), as well. For n numbers (Figure 3), it can be from $[0, 0.5)$ from 0 to n of them, with their occurrences being accounted for.

$|\{v_i, v_i \in [0, 0.5), 1 \le i \le n\}|$	0	...	j	...	n
Occurrence	1	...	$n!/((n-j)! \cdot j!)$	...	1

Figure 3. Contingency of n consecutive drawings from $[0, 1)$.

According to the formula given in Figure 3, for n numbers to be drawn from $[0, 1)$, a multiple of $n + 1$ drawings must be made in order to maintain the uniformity of distribution (w from Figure 1 becomes $n + 1$). In each of those drawings, we actually only pick one of n (random) numbers (from the $[0, 1)$ interval) as independent. In the $(j + 1)$-th drawing, the first j of them are to be from $[0, 0.5)$, while the rest are to be from $[0.5, 1)$. The algorithm implementing this strategy is given as Algorithm 1.

Algorithm 1 is ready to be used to calculate any OS (including the TS first reported here). For each sample drawn from the $U(0, 1)$ distribution (the array v in Algorithm 1), the output of it (the array u and its associated frequencies $n!/j!/(n - j)!$) can be modified to produce less information and operations (Algorithm 2). Calculation of the OS (OS_j output value in Algorithm 2) can be made to any precision, but for storing the result, a *single* data type (4 bytes) is enough (providing seven significant digits as the precision of the observed CDF of the OS). Along with a *byte* data type (j output value in Algorithm 2) to store each sampled OS, 5 bytes of memory is required, and the calculation of

$n!/(n-j)!/j!$ can be made at a later time, or can be tabulated in a separate array, ready to be used at a later time.

Algorithm 1: Balancing the drawings from uniform $U(0,1)$ distribution.

Input data: n ($2 \leq$ n, integer)
Steps:
   For i from 1 to n do v[i] ← Rand
   For j from 0 to n do
      For i from 1 to j do u[i] ← v[i]/2
      For i from j+1 to n do u[i] ← v[i]/2+1/2
      occ ← n!/j!/(n-j)!
      Output u[1], ..., u[n], occ
   EndFor
Output data: (n+1) samples (u) of sample size (n) and their occurrences (occ)

Algorithm 2: Sampling an order statistic (OS).

Input data: n ($2 \leq$ n, integer)
Steps:
   For i from 1 to n do v[i] ← Rand
   For j from 0 to n do
      For i from 1 to j do u[i] ← v[i]/2
      For i from j+1 to n do u[i] ← v[i]/2+1/2
      OS_j ← any Equations (5)–(12) with p_1 ← u[1], ..., p_n ← u[n]
      Output OS_j, j
   EndFor
Output data: (n+1) OS and their occurrences

As given in Algorithm 2, each use of the algorithm sampling OS will produce two associated arrays: OS_j (single data type) and j (byte data type); each of them with $n+1$ values. Running the algorithm r0 times will require $5 \cdot (n+1) \cdot r0$ bytes for storage of the results and will produce $(n+1) \cdot r0$ OSs, ready to be sorted (see Figure 1). With a large amount of internal memory (such as 64 GB when running on a 16/24 cores 64 bit computers), a single process can dynamically address very large arrays and thus can provide a good quality, sampled OS. To do this, some implementation tricks are needed (see Table 1).

Table 1. Software implementation peculiarities of MC simulation.

Constant/Variable/Type Value	Meaning
stt ← record v:single; c:byte; end	(OS_j, j) pair from Algorithm 2 stored in 5 bytes
mem ← 12,800,000,000	in bytes, 5*mem ← 64Gb, hardware limit
buf ← 1,000,000	the size of a static buffer of data (5*buf bytes)
stst ← array[0..buf-1] of stt	static buffer of data
dyst ← array of stst	dynamic array of buffers
lvl ← 1000	lvl + 1: number of points in the grid (see Figure 1)

Depending on the value of the sample size (n), the number of repetitions (r2) for sampling of OS, using Algorithm 2, from r0 ← $mem/(n+1)$ runs, is r2 ← $r0 \cdot (n+1)$, while the length (sts) of the variable (CDFst) storing the dynamic array (dyst) from Table 1 is sts ← $1 + r2/buf$. After sorting the OSs (of sttype, see Table 1; total number of r2) another trick is to extract a sample series at evenly spaced probabilities from it (from $InvCDF_0$ to $InvCDF_{1000}$ in Figure 1). For each pair in the sample (lvli varying from 0 to lvl = 1000 in Table 1), a value of the OS is extracted from CDFst array (which contains ordered

OS values and frequencies indexed from 0 to r2−1), while the MC-simulated population size is $r0 \cdot 2^n$. A program implementing this strategy is available upon request (*project_OS.pas*).

The associated objective (with any statistic) is to obtain its CDF and thus, by evaluating the CDF for the statistical value obtained from the sample, Equations (5)–(12), to associate a likelihood for the sampling. Please note that only in the lucky cases is it possible to do this; in the general case, only critical values (values corresponding to certain risks of being in error) or approximation formulas are available (see for instance [1–3,5,7–9]). When a closed form or an approximation formula is assessed against the observed values from an MC simulation (such as the one given in Table 1), a measure of the departure such as the standard error (SE) indicates the degree of agreement between the two. If a series of evenly spaced points ($lvl + 1$ points indexed from 0 to lvl in Table 1) is used, then a standard error of the agreement for inner points of it (from 1 to $lvl - 1$, see Equation (13)) is safe to be computed (where p_i stands for the observed probability while $\hat{p}_i$ for the estimated one).

$$SE = \sqrt{\frac{SS}{lvl - 1}}, \quad SS = \sum_{i=1}^{lvl-1} (p_i - \hat{p}_i)^2. \tag{13}$$

In the case of $lvl + 1$, evenly spaced points in the interval $[0, 1]$ in the context of MC simulation (as the one given in Table 1) providing the values of OS statistic in those points (see Figure 1), the observed cumulative probability should (and is) taken as $p_i = i/lvl$, while $\hat{p}_i$ is to be (and were) taken from any closed form or approximation formula for the CDF statistic (labeled $\hat{p}$) as $\hat{p}_i = \hat{p}(\text{InvCDF}_i)$, where InvCDF$_i$ are the values collected by the strategy given in Figure 1 operating on the values provided by Algorithm 2. Before giving a closed form for CDF of TS (Equation (12)) and proposing approximation formulas, other theoretical considerations are needed.

2.2. Further Theoretical Considerations Required for the Study

When the PDF is known, it does not necessarily imply that its statistical parameters $((\pi_j)_{1 \leq j \leq m}$ in Equations (1)–(3)) are known, and here, a complex problem of estimating the parameters of the population distribution from the sample (it then uses the same information as the one used to assess the quality of sampling) or from something else (and then it does not use the same information as the one used to assess the quality of sampling) can be (re)opened, but this matter is outside the scope of this paper.

The estimation of distribution parameters $(\pi_j)_{1 \leq j \leq m}$ for the data is, generally, biased by the presence of extreme values in the data, and thus, identifying the outliers along with the estimation of parameters for the distribution is a difficult task operating on two statistical hypotheses. Under this state of facts, the use of a hybrid statistic, such as the proposed one in Equation (12), seems justified. However, since the practical use of the proposed statistics almost always requires estimation of the population parameters (and in the examples given below, as well), a certain perspective on estimation methods is required.

Assuming that the parameters are obtained using the maximum likelihood estimation method (MLE, Equation (14); see [19]), one could say that the uncertainty accompanying this estimation is propagated to the process of detecting the outliers. With a series of τ statistics ($\tau = 6$ for Equations (5)–(10) and $\tau = 8$ for Equations (5)–(12)) assessing independently the risk of being in error (let be $\alpha_1, ..., \alpha_\tau$ those risks), assuming that the sample was drawn from the population, the unlikeliness of the event (α_{FCS} in Equation (15) below) can be ascertained safely by using a modified form of Fisher's "combining probability from independent tests" method (FCS, see [10,20,21]; Equation (15)), where CDF$_{\chi^2}(x; \tau)$ is the CDF of χ^2 distribution with τ degrees of freedom.

$$\max\left(\prod_{1 \leq i \leq n} \text{PDF}(x_i; (\pi_j)_{1 \leq j \leq m})\right) \to \min\left(\sum_{1 \leq j \leq m} \ln\left(\text{PDF}(x_i; (\pi_j)_{1 \leq j \leq m})\right)\right) \tag{14}$$

$$FCS = -\ln\left(\prod_{1\leq k\leq \tau} \alpha_k\right), \alpha_{FCS} = 1 - CDF_{\chi^2}(FCS; \tau). \tag{15}$$

Two known symmetrical distributions were used (PDF, see Equation (1)) to express the relative deviation from the observed distribution: Gauss (G2 in Equation (16)) and generalized Gauss–Laplace (GL in Equation (17)), where (in both Equations (16) and (17)) $z = (x - \mu)/\sigma$.

$$G2(x; \mu, \sigma) = (2\pi)^{-1/2} \sigma^{-1} e^{-z^2/2} \tag{16}$$

$$GL(x; \mu, \sigma, \kappa) = \frac{c_1}{\sigma} e^{-|c_0 z|^\kappa}, c_0 = \left(\frac{\Gamma(3/\kappa)}{\Gamma(1/\kappa)}\right)^{1/2}, c_1 = \frac{\kappa c_0}{2\Gamma(1/\kappa)}. \tag{17}$$

The distributions given in Equations (16) and (17) will be later used to approximate the CDF of TS as well as in the case studies of using the order statistics. For a sum ($x \leftarrow p_1 + ... + p_n$ in Equation (18)) of uniformly distributed ($p_1, ..., p_n \in U(0,1)$) deviates (as $\{p_1, ..., p_n\}$ in Equation (2)) the literature reports the Irwin–Hall distribution [22,23]. The $CDF_{IH}(x; n)$ is:

$$CDF_{IH}(x; n) = \sum_{k=0}^{\lfloor x \rfloor} (-1)^k \frac{(x-k)^n}{k!(n-k)!}. \tag{18}$$

3. Results and Discussion

3.1. The Analytical Formula of CDF for TS

The CDF of TS depends (only) on the sample size (n), e.g., $CDF_{TS}(x; n)$. As the proposed equation, Equation (12), resembles (as an inverse of) a sum of normal deviates, we expected that the CDF_{TS} will also be connected with the Irwin–Hall distribution, Equation (18). Indeed, the conducted study has shown that the inverse ($y \leftarrow 1/x$) of the variable (x) following the TS follows a distribution (1/TS) of which the CDF is given in Equation (19). Please note that the similarity between Equations (18) and (19) is not totally coincidental; 1/TS (see Equation (12)) is more or less a sum of uniform distributed deviates divided by the highest one. Also, for any positive arbitrary generated series, its ascending (x) and descending ($1/x$) sorts are complementary. With the proper substitution, $CDF_{1/TS}(y; n)$ can be expressed as a function of CDF_{IH}—see Equation (20).

$$CDF_{1/TS}(y; n) = \sum_{k=0}^{\lfloor n-y \rfloor} (-1)^k \frac{(n-y-k)^{n-1}}{k!(n-1-k)!} \tag{19}$$

$$CDF_{1/TS}(y; n) = CDF_{IH}(n - y; n - 1). \tag{20}$$

Unfortunately, the formulas, Equation (18) to Equation (20), are not appropriate for large n and p ($p = CDF_{1/TS}(y; n)$ from Equation (19)), due to the error propagated from a large number of numerical operations (see further Table 2 in Section 3.2). Therefore, for $p > 0.5$, a similar expression providing the value for $\alpha = 1 - p$ is more suitable. It is possible to use a closed analytical formula for $\alpha = 1 - CDF_{1/TS}(y; n)$ as well, Equation (21). Equation (21) resembles the Irwin–Hall distribution even more closely than Equation (20)—see Equation (22).

$$1 - CDF_{1/TS}(y; n) = \sum_{k=0}^{\lfloor y \rfloor - 1} (-1)^k \frac{(y-1-k)^n}{k!(n-1-k)!} \tag{21}$$

$$1 - CDF_{1/TS}(y; n) = CDF_{IH}(y - 1; n - 1). \tag{22}$$

For consistency in the following notations, one should remember the definition of CDF, see Equation (1), and then we mark the connection between notations in terms of the analytical expressions of the functions, Equation (23):

$$\text{CDF}_{TS}(x;n)=1-\text{CDF}_{1/TS}(1/x;n), \text{CDF}_{TS}(1/x;n)=1-\text{CDF}_{1/TS}(x;n),$$
$$\text{since } \text{InvCDF}_{TS}(p;n)\cdot\text{InvCDF}_{1/TS}(p;n)=1. \quad (23)$$

One should notice (Equation (1); Equation (23)) that the infimum for the domain of $1/TS$ (1) is the supremum for the domain of TS (1) and the supremum (n) for the domain of $1/TS$ is the infimum ($1/n$) for the domain of TS. Also, TS has the median ($p = \alpha = 0.5$) at $2/(n+1)$, while $1/TS$ has the median (which is also the mean and mode) at $(n+1)/2$. The distribution of $1/TS$ is symmetrical.

For $n = 2$, the $p = \text{CDF}_{1/TS}(y;n)$ is linear ($y + p = 2$), while for $n = 3$, it is a mixture of two square functions: $2p = (3-y)^2$, for $p \leq 0.5$ (and $y \geq 2$), and $2p + (y-1)^2 = 1$ for $p \geq 0.5$ (and $x \leq 2$). With the increase of n, the number of mixed polynomials of increasing degree defining its expression increases. Therefore, it has no way to provide an analytical expression for InvCDF of $1/TS$, not even for certain p values (such as 'critical' analytical functions).

The distribution of $1/TS$ can be further characterized by its central moments (Mean μ, Variance σ^2, Skewness γ_1, and Kurtosis κ in Equation (24)), which are closely connected with the Irwin–Hall distribution.

$$\text{For } 1/TS(y;n): \mu=(n+1)/2; \sigma^2=(n-1)/12, \gamma_1=0; \kappa=3-6/(5n-5). \quad (24)$$

3.2. Computations for the CDF of TS and Its Analytical Formula

Before we proceed in providing the simulation results, some computational issues must be addressed. Any of the formulas provided for CDF of TS (Equations (19) and (21); or Equations (20) and (22) both connected with Equation (18)), will provide almost exact calculations as long as computations with the formulas are conducted with an engine or package that performs the operations with rational numbers to an infinite precision (such as is available in the Mathematica software [24]), when also the value of y ($y \leftarrow 1/x$, of floating point type) is converted to a rounded, rational number. Otherwise, with increasing n, the evaluation of CDF for TS using either Equation (19) to Equation (22) carries huge computational errors (see the alternating sign of the terms in the sums of Equations (18), (19), and (21)). In order to account for those computational errors (and to reduce their magnitude) an alternate formula for the CDF of TS is proposed (Algorithm 3), combining the formulas from Equations (19) and (21), and reducing the number of summed terms.

Algorithm 3: Avoiding computational errors for TS.

Input data: n (n ≥ 2, integer), x (1 ≤ x ≤ 1/n, real number, double precision)
$y \leftarrow 1/x;$ $//p_{1/TS} \leftarrow$ Equation (19), $\alpha_{1/TS} \leftarrow$ Equation (21)
if y <(n+1)/2
$\quad p \leftarrow \sum_{k=0}^{\lfloor y \rfloor -1}(-1)^k \frac{(y-1-k)^n}{k!(n-1-k)!}; \alpha \leftarrow 1-p$
else if y >(n+1)/2
$\quad \alpha \leftarrow \sum_{k=0}^{\lfloor n-y \rfloor}(-1)^k \frac{(n-y-k)^{n-1}}{k!(n-1-k)!}; p \leftarrow 1-\alpha$
else
$\quad \alpha \leftarrow 0.5; p \leftarrow 0.5$
Output data: $\alpha = \alpha_{1/TS} = p_{TS} \leftarrow \text{CDF}_{TS}(x;n)$ and $p = p_{1/TS} = \alpha_{TS} \leftarrow 1 - p_{TS}$

Table 2 contains the sums of the residuals ($SS = \sum_{i=1}^{999}(p_i - \hat{p}_i)^2$ in Equation (13), $lvl = 1000$) of the agreement between the observed CDF of TS ($p_i = i/1000$, for i from 1 to 999) and the calculated CDF of TS (the $\hat{p}_i$ values are calculated using Algorithm 3 from $x_i = \mathrm{InvCDF}(i/1000; n)$ for i from 1 to 999) for some values of the sample size (n). To prove the previous given statements, Table 2 provides the square sums of residuals computed using three alternate formulas (from Equation (20) and from Equation (22), along with the ones from Algorithm 3).

Table 2. Square sums of residuals calculated in double precision (IEEE 754 binary64, 64 bits).

n	p_i Calculated with Equation (19)	p_i Calculated with Equation (21)	p_i Calculated with Algorithm 4
34	$3.0601572482628 \times 10^{-8}$	$3.0601603616294 \times 10^{-8}$	$3.0601364353173 \times 10^{-8}$
35	$6.0059397209079 \times 10^{-8}$	$6.0057955311142 \times 10^{-8}$	$6.0057052975471 \times 10^{-8}$
36	$1.1567997676343 \times 10^{-8}$	$1.1572997605838 \times 10^{-8}$	$1.1567370749831 \times 10^{-8}$
37	$8.9214456109544 \times 10^{-8}$	$8.9215230398577 \times 10^{-8}$	$8.9213063043724 \times 10^{-8}$
38	$1.1684682533384 \times 10^{-8}$	$1.1681544866285 \times 10^{-8}$	$1.1677646550768 \times 10^{-8}$
39	$1.2101651325053 \times 10^{-8}$	$1.2181659126285 \times 10^{-8}$	$1.2100378665608 \times 10^{-8}$
40	$1.1041708665520 \times 10^{-7}$	$1.1043952711846 \times 10^{-7}$	$1.1036003349029 \times 10^{-7}$
41	$7.2871410520319 \times 10^{-8}$	$7.2755412302319 \times 10^{-8}$	$7.2487977100103 \times 10^{-8}$
42	$1.9483807018501 \times 10^{-8}$	$1.9626447735907 \times 10^{-8}$	$1.9273186509959 \times 10^{-8}$
43	$3.1128379331196 \times 10^{-8}$	$1.7088238120170 \times 10^{-8}$	$1.3899520242290 \times 10^{-8}$
44	$8.7810761126831 \times 10^{-8}$	$3.8671367222236 \times 10^{-8}$	$1.0878689813951 \times 10^{-8}$
45	$1.1914784602127 \times 10^{-7}$	$3.1416715528555 \times 10^{-7}$	$5.8339481916925 \times 10^{-8}$
46	$2.0770754629042 \times 10^{-6}$	$1.2401177918843 \times 10^{-6}$	$4.4594953399233 \times 10^{-8}$
47	$5.0816356972050 \times 10^{-7}$	$4.1644326761832 \times 10^{-7}$	$1.8942487765410 \times 10^{-8}$
48	$1.5504732794049 \times 10^{-6}$	$5.5760558048026 \times 10^{-6}$	$5.7292512517324 \times 10^{-8}$
49	$1.1594466754136 \times 10^{-5}$	$6.4164330856396 \times 10^{-6}$	$1.7286761495408 \times 10^{-7}$
50	$1.0902858025759 \times 10^{-5}$	$8.0190771776360 \times 10^{-6}$	$8.5891058550425 \times 10^{-8}$
51	$6.4572577668164 \times 10^{-6}$	$1.6023753568028 \times 10^{-4}$	$1.9676739380922 \times 10^{-8}$
52	$1.0080944275181 \times 10^{-4}$	$9.1080176774820 \times 10^{-5}$	$1.0359121739272 \times 10^{-7}$
53	$9.3219609856284 \times 10^{-4}$	$2.7347575817507 \times 10^{-4}$	$1.5873847007230 \times 10^{-8}$
54	$4.8555844748161 \times 10^{-4}$	$1.6086902937472 \times 10^{-3}$	$9.2930071189138 \times 10^{-9}$
55	$6.2446720485774 \times 10^{-4}$	$1.6579954395873 \times 10^{-3}$	$1.2848119194342 \times 10^{-7}$

In red: computing affected digits.

As given in Table 2, the computational errors by using either Equation (20) (or Equation (19)) and Equation (22) (or Equation (21)) until $n = 34$ are reasonably low, while from $n = 42$, they become significant. As can be seen (red values in Table 2), double precision alone cannot cope with the large number of computations, especially as the terms in the sums are constantly changing their signs (see $(-1)^k$ in Equations (19) and (21)).

The computational errors using Algorithm 3 are reasonably low for the whole domain of the simulated CDF of TS (with n from 2 to 55), but the combined formula (Algorithm 3) is expected to lose its precision for large n values, and therefore, a solution to safely compute (CDF for IH, TS and $1/TS$) is to operate with rational numbers.

One other alternative is to use GNU GMP (Multiple Precision Arithmetic Library [25]). The calculations are the same (Algorithm 3); the only difference is the way in which the temporary variables are declared (instead of *double*, the variables become mpf_t initialized later with a desired precision).

For convenience, the FreePascal [26] implementation for CDF of the Irwin–Hall distribution (Equation (18), called in the context of evaluating the CDF of TS in Equations (20) and (22)) is given as Algorithm 4.

Algorithm 4: FreePascal implementation for calculating the CDF of *IH*.

Input data: n (integer), x (real number, double precision);
var k,i: integer; //*integer* enough for n < 32,768
var z,y: mpf_t; //*double* or *extended* instead of *mpf_t*
Begin //CDF for Irwin–Hall distribution
  mpf_set_default_prec(128); //or bigger, 256, 512, ...
  mpf_init(y); mpf_init(z); //y := 0.0;
  for k := trunc(x) downto 0 do **begin** //main loop
    If(k mod 2 = 0) // z := 1.0 or z := −1.0;
      then mpf_set_si(z,1) //z := 1.0;
      else mpf_set_si(z,-1); //z := -1.0;
    for i := n − k downto 1 do z := z*(x − k)/i;
    for i := k downto 1 do z := z*(x− k)/i;
    y := y + z;
  end;
  pIH_gmp := mpf_get_d(y); mpf_clear(z); mpf_clear(y);
End;
Output data: p (real number, double precision)

In Algorithm 4, the changes made to a classical code running without GNU GMP floating point arithmetic functions are written in blue color. For convenience, the combined formula (Algorithm 3) trick for avoiding the computation errors can be implemented with the code given as Algorithm 4 at the call level, Equation (25). If `pIH(x:double; n:integer):double` returns the value from Algorithm 4, then $pg1$, as given in Equation (25), safely returns the combined formula (Algorithm 3) with (or without) GNU GMP.

$$pg1 \leftarrow \begin{cases} 1 - pIH(n-1, n-1/x), & \text{if } x(n+1) < 2. \\ pIH(n-1, 1/x - 1), & \text{otherwise.} \end{cases} \quad (25)$$

Regarding Table 2, Algorithm 4 listed data, from $n = 2$ to $n = 55$, the calculation of the residuals were made with *double* (64 bits), *extended* (FreePascal 80 bits), and *mpf_t*-128 bits (GNU GMP). The sum of residuals (for all n from 2 to 55) differs from *double* to *extended* with less than 10^{-11} and the same for *mpf_t* with 128 bits, which safely provides confidence in the results provided in Table 2 for the combined formula (last column, Algorithm 4). The deviates for agreement in the calculation of CDF for *TS* are statistically characterized by *SE* (Equation (13)), *min*, and *max* in Table 3.

The *SE* of agreement (Table 3) between the expected value and the observed one (Algorithm 4, Equation (12), Table 1) of the $CDF_{1/TS}(x;n)$ is safely below the resolution for the grid of observing points ($lvl^{-1} = 10^{-3}$ in Table 1; $SE \leq 1.2 \times 10^{-5}$ in Table 3; two orders of magnitude). By using Algorithm 4, Figures 4–7 depict the shapes of $CDF_{TS}(x;n)$, $CDF_{1/TS}(x;n)$, $InvCDF_{TS}(x;n)$, and $InvCDF_{1/TS}(x;n)$ for n from 2 to 20.

Finally, for the domain of the simulated CDF of the *TS* population for n from 2 to 54, the error in the odd points of the grid (for $1000 \cdot p$ from 1 to 999 with a step of 2) is depicted in Figure 8 (the calculations of theoretical CDF for *TS* made with *gmpfloat* at a precision of at least 256 bits). As can be observed in Figure 8, the difference between p and $\hat{p}$ is rarely larger than 10^{-5} and never larger than 3×10^{-5} (the boundary of the representation in Figure 8) for n ranging from 2 to 54.

Table 3. Descriptive for the agreement in the calculation of the CDF of TS (Equation (12) vs. Algorithm 4).

n	SE	minep	maxep	n	SE	minep	maxep	n	SE	minep	maxep
2	3.0×10^{-6}	-2.1×10^{-6}	1.8×10^{-6}	20	5.4×10^{-6}	-4.1×10^{-6}	3.9×10^{-6}	38	3.4×10^{-6}	-7.3×10^{-6}	6.1×10^{-6}
3	3.2×10^{-6}	-2.4×10^{-6}	2.7×10^{-6}	21	3.0×10^{-6}	-4.5×10^{-6}	4.1×10^{-6}	39	3.5×10^{-6}	-7.3×10^{-6}	6.4×10^{-6}
4	3.5×10^{-6}	-2.3×10^{-6}	2.7×10^{-6}	22	6.3×10^{-6}	-4.8×10^{-6}	4.0×10^{-6}	40	1.1×10^{-5}	-7.2×10^{-6}	5.5×10^{-6}
5	4.2×10^{-6}	-2.8×10^{-6}	2.2×10^{-6}	23	5.6×10^{-6}	-5.6×10^{-6}	4.6×10^{-6}	41	8.5×10^{-6}	-7.2×10^{-6}	7.4×10^{-6}
6	2.8×10^{-6}	-3.2×10^{-6}	2.4×10^{-6}	24	4.0×10^{-6}	-6.4×10^{-6}	4.6×10^{-6}	42	4.4×10^{-6}	-7.0×10^{-6}	7.8×10^{-6}
7	4.4×10^{-6}	-3.3×10^{-6}	3.1×10^{-6}	25	4.1×10^{-6}	-6.3×10^{-6}	4.5×10^{-6}	43	3.7×10^{-6}	-6.5×10^{-6}	6.9×10^{-6}
8	3.5×10^{-6}	-3.7×10^{-6}	2.6×10^{-6}	26	1.2×10^{-5}	-6.2×10^{-6}	5.1×10^{-6}	44	3.3×10^{-6}	-6.1×10^{-6}	7.0×10^{-6}
9	3.7×10^{-6}	-3.9×10^{-6}	2.2×10^{-6}	27	1.2×10^{-5}	-6.3×10^{-6}	4.9×10^{-6}	45	7.6×10^{-6}	-6.1×10^{-6}	6.8×10^{-6}
10	4.5×10^{-6}	-3.7×10^{-6}	2.9×10^{-6}	28	7.8×10^{-6}	-6.3×10^{-6}	5.1×10^{-6}	46	6.7×10^{-6}	-6.1×10^{-6}	6.9×10^{-6}
11	5.7×10^{-6}	-3.7×10^{-6}	2.7×10^{-6}	29	7.2×10^{-6}	-6.6×10^{-6}	5.4×10^{-6}	47	4.4×10^{-6}	-6.2×10^{-6}	7.3×10^{-6}
12	7.6×10^{-6}	-3.9×10^{-6}	2.5×10^{-6}	30	3.5×10^{-6}	-6.3×10^{-6}	5.7×10^{-6}	48	7.6×10^{-6}	-6.2×10^{-6}	8.0×10^{-6}
13	5.2×10^{-6}	-3.8×10^{-6}	3.0×10^{-6}	31	4.1×10^{-6}	-6.2×10^{-6}	5.0×10^{-6}	49	1.3×10^{-5}	-6.3×10^{-6}	7.8×10^{-6}
14	5.6×10^{-6}	-4.3×10^{-6}	3.2×10^{-6}	32	5.2×10^{-6}	-6.0×10^{-6}	4.9×10^{-6}	50	9.3×10^{-6}	-6.0×10^{-6}	7.0×10^{-6}
15	1.0×10^{-5}	-3.8×10^{-6}	3.5×10^{-6}	33	3.5×10^{-6}	-6.0×10^{-6}	4.5×10^{-6}	51	4.4×10^{-6}	-6.4×10^{-6}	7.0×10^{-6}
16	6.9×10^{-6}	-3.9×10^{-6}	3.6×10^{-6}	34	5.5×10^{-6}	-6.6×10^{-6}	4.3×10^{-6}	52	1.0×10^{-5}	-6.4×10^{-6}	6.4×10^{-6}
17	8.4×10^{-6}	-4.2×10^{-6}	3.5×10^{-6}	35	7.8×10^{-6}	-6.3×10^{-6}	5.2×10^{-6}	53	4.0×10^{-6}	-6.1×10^{-6}	6.1×10^{-6}
18	5.1×10^{-6}	-4.1×10^{-6}	4.1×10^{-6}	36	3.4×10^{-6}	-6.7×10^{-6}	5.7×10^{-6}	54	3.1×10^{-6}	-6.4×10^{-6}	6.7×10^{-6}
19	5.4×10^{-6}	-4.2×10^{-6}	4.4×10^{-6}	37	9.4×10^{-6}	-6.8×10^{-6}	6.4×10^{-6}	55	1.1×10^{-5}	-6.7×10^{-6}	7.1×10^{-6}

$minep = min(p_i - \hat{p}_i)$, $maxep = max(p_i - \hat{p}_i)$.

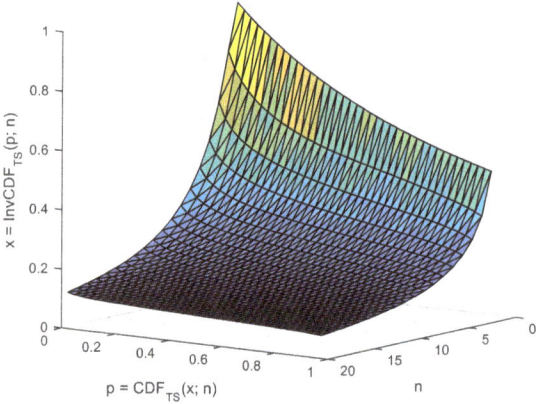

Figure 4. $\text{InvCDF}_{TS}(x; n)$ for $n = 2$ to 20.

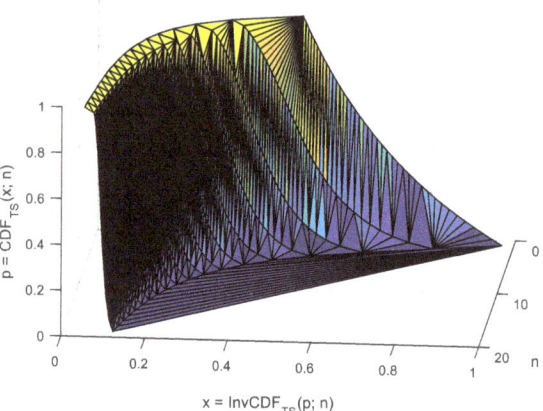

Figure 5. $\text{CDF}_{TS}(x; n)$ for $n = 2$ to 20.

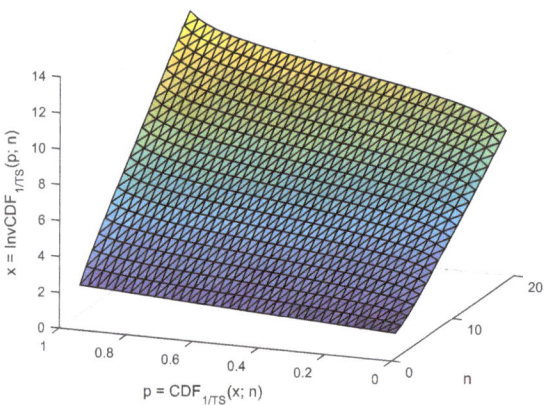

Figure 6. $\text{InvCDF}_{1/TS}(x; n)$ for $n = 2$ to 20.

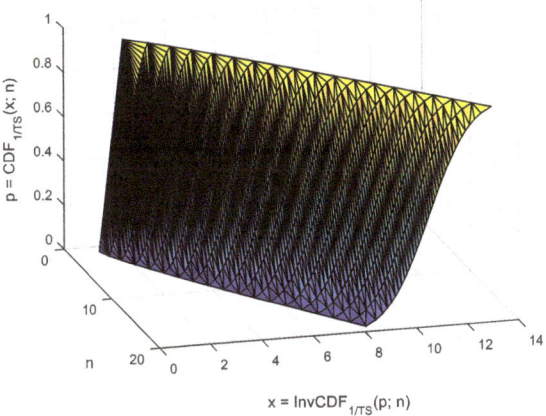

Figure 7. $CDF_{1/TS}(x;n)$ for $n = 2$ to 20.

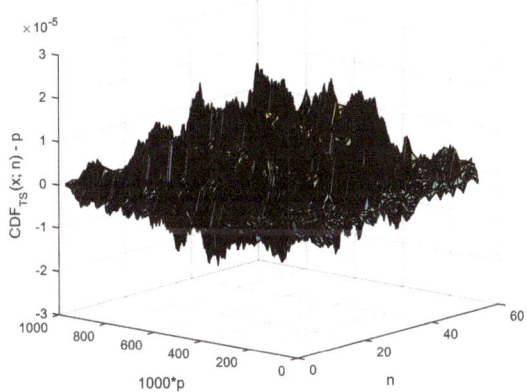

Figure 8. Agreement estimating CDF_{TS} for $n = 2...54$ and $1000p = 1...999$ with a step of 2.

Based on the provided results, one may say that there is no error in saying that Equations (19) and (21) are complements (see Equation (23) as well) of the CDF of TS given as Equation (12). As long as the calculations (of either Equations (19) and (21)) are conducted using rational numbers, either formula provides the most accurate result. The remaining concerns are how large those numbers can be (e.g., the range of n). This is limited only by the amount of memory available and how precise the calculations are. This reaches the maximum defined by the measurement of data precision, and finally, the resolutions are provided, which are given by the precision of converting (if necessary) the TS value given by Equation (12) from float to rational. Either way, some applications prefer approximate formulas, which are easier to calculate, and are considered common knowledge for interpreting the results. For those reasons, the next section describes approximation formulas.

3.3. Approximations of CDF of TS with Known Functions

Considering, once again, Equation (24), for sufficiently large n, the distribution of $1/TS$ is approximately normal (Equation (26). For normal Gauss distribution, see Equation (16)).

$$PDF_{1/TS}(y;n) \xrightarrow{n \to \infty} PDF_{G2}((n+1)/2; \sqrt{(n-1)/12}). \tag{26}$$

Even better (than Equation (26)), for large values of n, a generalized Gauss–Laplace distribution (see Equation (17)) can be used to approximate the $1/TS$ statistic. Furthermore, for those looking for critical values of the TS statistic, the approximation of the $1/TS$ statistic to a generalized Gauss–Laplace distribution may provide safe critical values for large n. One way to derive the parameters of the generalized Gauss–Laplace distribution approximating the $1/TS$ statistic is by connecting the kurtosis and skewness of the two (Equation (27)).

$$Ku(\beta) = \frac{\Gamma(\frac{5}{\beta})\Gamma(\frac{1}{\beta})}{\Gamma(\frac{3}{\beta})\Gamma(\frac{3}{\beta})} \to \beta = Ku^{-1}\left(3 - \frac{6}{5n-5}\right),\ \alpha = \sqrt{\frac{n-1}{12}\frac{\Gamma(1/\beta)}{\Gamma(3/\beta)}}. \tag{27}$$

With α and β given by Equation (27) and $\mu = (n+1)/2$ (Equation (24)), the PDF of the generalized Gauss–Laplace distribution (Equation (17)), which approximates $1/TS$ (for large n), is given in Equation (28).

$$\text{PDF}_{GL}(x;\mu,\alpha,\beta) = \frac{\beta}{2\alpha\Gamma(1/\beta)} e^{-\left(\frac{|x-\mu|}{\alpha}\right)^{\beta}}. \tag{28}$$

The errors of approximation (with Equation (29)) of $p_i = \text{CDF}_{1/TS}$ (from Algorithm 3) with $\hat{p}_i = \text{CDF}_{GL}$ (from Equations (27) and 28) are depicted in Figure 9 using a grid of 52×999 points for $n = 50...101$ and $p = 0.001...0.999$.

$$SE = \sqrt{\sum_{i=1}^{999} \frac{(p_i - \hat{p}_i)^2}{999}},\ p_i = \frac{i}{10^3},\ \hat{p}_i = \text{CDF}_{GL}(\text{InvCDF}_{1/TS}(p_i;n);\alpha,\beta). \tag{29}$$

As can be observed in Figure 9, the confidence in approximation of $1/TS$ with the GL increases with the sample size (n), but the increase is less than linear. The tendency is to approximately linearly decrease with an exponential increase.

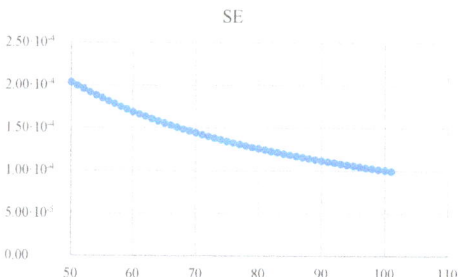

Figure 9. Standard errors (SE) as function of sample size (n) for the approximation of $1/TS$ with GL (Equation (29)).

The calculation of CDF for $1/TS$ is a little tricky, as anticipated previously (see Section 3.2). To avoid the computation errors in the calculation of CDF_{TS}, a combined formula is more appropriate (Algorithms 3 and 4). With $p_{1/TS} \leftarrow \text{CDF}_{1/TS}(y;n)$ and $\alpha_{1/TS} \leftarrow 1 - \text{CDF}_{1/TS}(y;n)$, depending on the value of y ($y \leftarrow 1/x$, where x is the sample statistic of TS, Equation (12)), only one (from α and p, where $\alpha + p = 1$) is suitable for a precise calculation.

An important remark at this point is that $(n+1)/2$ is the median, mean, and mode for $1/TS$ (see Section 3.1). Indeed, any symbolic calculation with either of the formulas from Equation (19) to Equation (22) will provide that $\text{CDF}_{1/TS}((n+1)/2;n) = 0.5$, or, expressed with InvCDF, $\text{InvCDF}_{1/TS}(0.5;n) = (n+1)/2$.

3.4. The Use of CDF for TS to Measure the Departure between an Observed Distribution and a Theoretical One

With any of Equations (5)–(12), a likelihood to observe an observed sample can be ascertained. One may ask which statistic is to be trusted. The answer is, at the same time, none and all, as the problem of fitting the data to a certain distribution involves the estimation of the distribution's parameters—such as using MLE, Equation (14). In this process of estimation, there is an intrinsic variability that cannot be ascertained by one statistic alone. This is the reason that calculating the risk of being in error from a battery of statistics is necessary, Equation (15).

Also, one may say that the $g1$ statistic (Equation (11)) is not associated with the sample, but to its extreme value(s), while others may say the opposite. Again, the truth is that both are right, as in certain cases, samples containing outliers are considered not appropriate for the analysis [27], and in those cases, there are exactly two modes of action: to reject the sample or to remove the outlier(s). Figure 10 gives the proposed strategy of assessing the samples using order statistics.

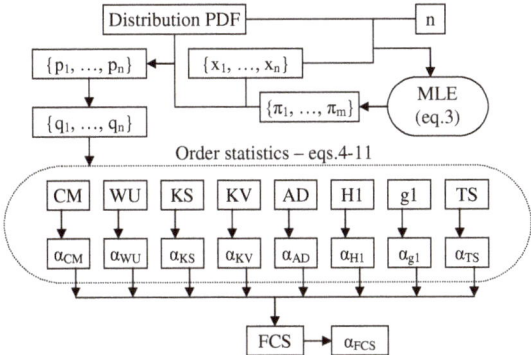

Figure 10. Using the order statistics to measure the likelihood of sampling.

As other authors have noted, in nonparametric problems, it is known that order statistics, i.e., the ordered set of values in a random sample from least to greatest, play a fundamental role. 'A considerable amount of new statistical inference theory can be established from order statistics assuming nothing stronger than continuity of the cumulative distribution function of the population' as [28] noted, a statement that is perfectly valid today.

In the following case studies, the values of the sample statistics were calculated with Equations (5)–(10) ($AD, KS, CM, KV, WU, H1$; see also Figure 10), while the risks of being in error—associated with the values of sample statistics ($\alpha_{Statistic}$ for those)—were calculated with the program developed and posted online available at http://l.academicdirect.org/Statistics/tests. The $g1_{Statistic}$ (Equation (11)) and α_{g1} were calculated as given in [15], while the $TS_{Statistic}$ (Equation (12)) was calculated with Algorithm 4. For FCS and α_{FCS}, Equation (15) was used.

Case study 1.

Data: "Example 1" in [29]; Distribution: Gauss (Equation (16)); Sample size: $n = 10$; Population parameters (MLE, Equation (14)): $\mu = 575.2$; $\sigma = 8.256$; Order statistics analysis is given in Table 4. Conclusion: at $\alpha = 5\%$ risk of being in error, the sample does not have an outlier ($\alpha_{g1} = 11.2\%$) but it is a bad drawing from normal (Gauss) distribution, with less than the imposed level ($\alpha = 5\%$) likelihood to appear from a random draw ($\alpha_{FCS} = 4.5\%$).

Table 4. Order statistics analysis for case studies 1 to 10.

Case	Parameter	AD	KS	CM	KV	WU	H1	g1	TS	FCS
1	Statistic	1.137	1.110	0.206	1.715	0.182	5.266	0.494	4.961	15.80
	$\alpha_{Statistic}$	0.288	0.132	0.259	0.028	0.049	0.343	0.112	0.270	0.045
2	Statistic	0.348	0.549	0.042	0.934	0.039	7.974	0.496	6.653	6.463
	$\alpha_{Statistic}$	0.894	0.884	0.927	0.814	0.844	0.264	0.109	0.107	0.596
3	Statistic	0.617	0.630	0.092	1.140	0.082	4.859	0.471	5.785	4.627
	$\alpha_{Statistic}$	0.619	0.742	0.635	0.486	0.401	0.609	0.451	0.627	0.797
4	Statistic	0.793	0.827	0.144	1.368	0.129	3.993	0.482	4.292	8.954
	$\alpha_{Statistic}$	0.482	0.420	0.414	0.190	0.154	0.524	0.255	0.395	0.346
5	Statistic	0.440	0.486	0.049	0.954	0.047	104.2	0.500	103.2	5.879
	$\alpha_{Statistic}$	0.810	0.963	0.884	0.850	0.742	0.359	0.034	0.533	0.661
6	Statistic	0.565	0.707	0.083	1.144	0.061	83.32	0.499	82.17	5.641
	$\alpha_{Statistic}$	0.683	0.675	0.673	0.578	0.580	0.455	0.247	0.305	0.687
7	Statistic	1.031	1.052	0.170	1.662	0.149	52.66	0.494	51.00	11.24
	$\alpha_{Statistic}$	0.320	0.202	0.333	0.067	0.106	0.471	0.729	0.249	0.188
8	Statistic	0.996	0.771	0.132	1.375	0.127	22.201	0.460	27.95	5.933
	$\alpha_{Statistic}$	0.322	0.556	0.451	0.248	0.162	0.853	0.980	0.978	0.655
9	Statistic	0.398	0.576	0.058	1.031	0.051	31.236	0.489	32.04	2.692
	$\alpha_{Statistic}$	0.853	0.869	0.828	0.728	0.694	0.577	0.746	0.507	0.952
10	Statistic	0.670	0.646	0.092	1.170	0.085	11.92	0.460	14.66	3.549
	$\alpha_{Statistic}$	0.583	0.753	0.627	0.488	0.373	0.747	0.874	0.879	0.895

Case study 2.

Data: "Example 3" in [29]; Distribution: Gauss (Equation (16)); Sample size: $n = 15$; Population parameters (MLE, Equation (14)): $\mu = 0.018$; $\sigma = 0.532$; Order statistics analysis is given in Table 4. Conclusion: at $\alpha = 5\%$ risk of being in error, the sample does not have an outlier ($\alpha_{g1} = 10.9\%$) and it is a good drawing from normal (Gauss) distribution, with more than the imposed level ($\alpha = 5\%$) likelihood to appear from a random draw ($\alpha_{FCS} = 59.6\%$).

Case study 3.

Data: "Example 4" in [29]; Distribution: Gauss (Equation (16)); Sample size: $n = 10$; Population parameters (MLE, Equation (14)): $\mu = 3.406$; $\sigma = 0.732$; Order statistics analysis is given in Table 4. Conclusion: at $\alpha = 5\%$ risk of being in error, the sample does not have an outlier ($\alpha_{g1} = 45.1\%$) and it is a good drawing from normal (Gauss) distribution, with more than the imposed level ($\alpha = 5\%$) likelihood to appear from a random draw ($\alpha_{FCS} = 79.7\%$).

Case study 4.

Data: "Example 5" in [29]; Distribution: Gauss (Equation (16)); Sample size: $n = 8$; Population parameters (MLE, Equation (14)): $\mu = 4715$; $\sigma = 140.8$; Order statistics analysis is given in Table 4. Conclusion: at $\alpha = 5\%$ risk of being in error, the sample does not have an outlier ($\alpha_{g1} = 25.5\%$) and it is a good drawing from normal (Gauss) distribution, with more than the imposed level ($\alpha = 5\%$) likelihood to appear from a random draw ($\alpha_{FCS} = 34.6\%$).

Case study 5.

Data: "Table 4" in [15]; Distribution: Gauss (Equation (16)); Sample size: $n = 206$; Population parameters (MLE, Equation (14)): $\mu = 6.481$; $\sigma = 0.829$; Order statistics analysis is given in Table 4. Conclusion: at $\alpha = 5\%$ risk of being in error, the sample have an outlier ($\alpha_{g1} = 3.4\%$) and it is a good drawing from normal (Gauss) distribution, with more than the imposed level ($\alpha = 5\%$) likelihood to appear from a random draw ($\alpha_{FCS} = 66.1\%$).

Case study 6.

Data: "Table 1, Column 1" in [30]; Distribution: Gauss (Equation (16)); Sample size: $n = 166$; Population parameters (MLE, Equation (14)): $\mu = -0.348$; $\sigma = 1.8015$; Order statistics analysis is given in Table 4. Conclusion: at $\alpha = 5\%$ risk of being in error, the sample does not have an outlier ($\alpha_{g1} = 24.7\%$) and it is a good drawing from normal (Gauss) distribution, with more than the imposed level ($\alpha = 5\%$) likelihood to appear from a random draw ($\alpha_{FCS} = 68.7\%$).

Case study 7.

Data: "Table 1, Set BBB" in [31]; Distribution: Gauss (Equation (16)); Sample size: $n = 105$; Population parameters (MLE, Equation (14)): $\mu = -0.094$; $\sigma = 0.762$; Order statistics analysis is given in Table 4. Conclusion: at $\alpha = 5\%$ risk of being in error, the sample does not have an outlier ($\alpha_{g1} = 72.9\%$) and it is a good drawing from normal (Gauss) distribution, with more than the imposed level ($\alpha = 5\%$) likelihood to appear from a random draw ($\alpha_{FCS} = 18.8\%$).

Case study 8.

Data: "Table 1, Set SASCAII" in [31]; Distribution: Gauss (Equation (16)); Sample size: $n = 47$; Population parameters (MLE, Equation (14)): $\mu = 1.749$; $\sigma = 0.505$; Order statistics analysis is given in Table 4. Conclusion: at $\alpha = 5\%$ risk of being in error, the sample does not have an outlier ($\alpha_{g1} = 98.0\%$) and it is a good drawing from normal (Gauss) distribution, with more than the imposed level ($\alpha = 5\%$) likelihood to appear from a random draw ($\alpha_{FCS} = 65.5\%$).

Case study 9.

Data: "Table 1, Set TaxoIA" in [31]; Distribution: Gauss (Equation (16)); Sample size: $n = 63$; Population parameters (MLE, Equation (14)): $\mu = 0.744$; $\sigma = 0.670$; Order statistics analysis is given in Table 4. Conclusion: at $\alpha = 5\%$ risk of being in error, the sample does not have an outlier ($\alpha_{g1} = 74.6\%$) and it is a good drawing from normal (Gauss) distribution, with more than the imposed level ($\alpha = 5\%$) likelihood to appear from a random draw ($\alpha_{FCS} = 95.2\%$).

Case study 10.

Data: "Table 1, Set ERBAT" in [31]; Distribution: Gauss (Equation (16)); Sample size: $n = 25$; Population parameters (MLE, Equation (14)): $\mu = 0.379$; $\sigma = 1.357$; Order statistics analysis is given in Table 4. Conclusion: at $\alpha = 5\%$ risk of being in error, the sample does not have an outlier ($\alpha_{g1} = 87.9\%$) and it is a good drawing from normal (Gauss) distribution, with more than the imposed level ($\alpha = 5\%$) likelihood to appear from a random draw ($\alpha_{FCS} = 89.5\%$).

3.5. The Patterns in the Order Statistics

A cluster analysis on the risks of being in error, provided by the series of order statistics on the case studies considered in this study, may reveal a series of peculiarities (Figures 11 and 12). The analysis given here is based on the series of the above given case studies in order to illustrate similarities (and not to provide a 'gold standard' as in [32] or in [33]).

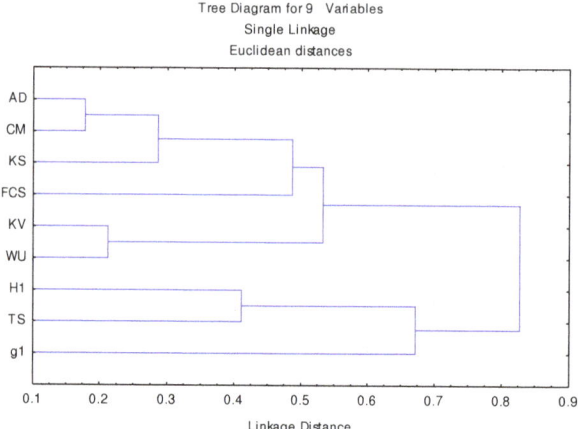

Figure 11. Euclidian distances between the risks being in error provided by the order statistics.

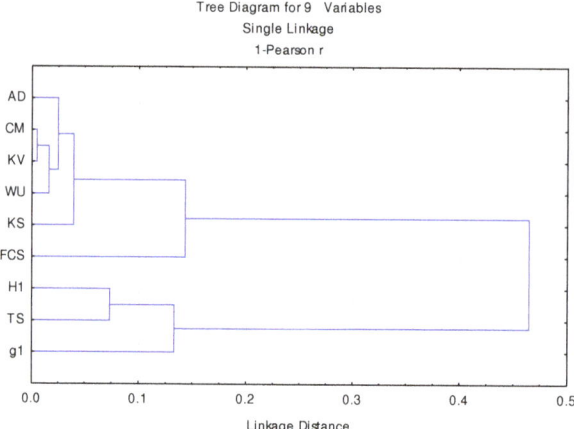

Figure 12. Pearson disagreement between the risks being in error provided by the order statistics.

Both clustering methods illustrated in Figures 11 and 12 reveal two distinct groups of statistics: {AD, CM, KV, WU, KS} and {H1, TS, g1}. The combined test FCS is also attracted (as expected) to the largest group. When looking at single Euclidean distances (Figure 11) of the largest group, two other associations should be noticed {AD, CM, KS} and {KV, WU}, suggesting that those groups carry similar information, but when looking at the Pearson disagreements (Figure 12), we must notice that the subgroups are changed {CM, KV, WU}, {AD}, and {KS}, with no hint of an association with their calculation formulas (Equations (5)–(9)); therefore, their independence should not be dismissed. The second group {H1, TS, g1} is more stable, maintaining the same clustering pattern of the group ({H1, TS}, {g1} in Figure 12).

Taking into account that the g1 test (Equation (11)) was specifically designed to account for outliers suggests that the H1 and TS tests are more sensitive to the outliers than other statistics, and therefore, when the outliers (or just the presence of extreme values) are the main concern in the sampling, it is strongly suggested to use those tests. The H1 statistic is a Shannon entropy formula applied in the probability space of the sample. When accounting for this aspect in the reasoning, the rassociation of the H1 with TS suggests that TS is a sort of entropic measure (max-entropy, to be more exact [34], a limit case of generalized Rényi's entropy [35]). Again, the g1 statistic is alone in this entropic group, suggesting that it carries a unique fingerprint about the sample—specifically, about its extreme value (see Equation (11))—while the others account for the context (the rest of the sampled values, Equations (10) and (12)).

Regarding the newly proposed statistic (TS), from the given case studies, the fact that it belongs to the {H1, TS, g1} group strongly suggests that it is more susceptible to the presence of outliers (such as g1, purely defined for this task, and unlike the well known statistics defined by Equations (5)–(9)).

Moreover, one may ask that, if based on the risks being in error provided by the statistics from case studies 1 to 10, some peculiarity about TS or another statistic involved in this study could be revealed. An alternative is to ask if the values of risks can be considered to be belonging to the same population or not, and for this, the K-sample Anderson–Darling test can be invoked [36]. With the series of probabilities, there are actually $2^9 - 1 - 9 = 502$ tests to be conducted (for each subgroup of 2, 3, 4, 5, 6, 7, 8, and 9 statistics picked from nine possible choices) and for each of them, the answer is same: At the 5% risk of being in error, it cannot be rejected that the groups (of statistics) were selected from identical populations (of statistics), so, overall, any of those statistics perform the same.

The proposed method may find its uses in testing symmetry [37], as a homogeneity test [38] and, of course, in the process of detecting outliers [39].

3.6. Another Rank Order Statics Method and Other Approaches

The series of rank order statistics included in this study, Equations (5)–(11), covers the most known rank order statistics reported to date. However, when considering a new order statistic not included there, the use of it in the context of combining methods, Equation (15), only increases the degrees of freedom τ, while the design of using (Figure 10) is changed accordingly.

It should be noted that the proposed approach is intended to be used for small sample sizes, when no statistic alone is capable of high precision and high trueness. With the increasing sample size, all statistics should converge to the same risk of being in error and present other alternatives, such as the superstatistical approach [40]. In the same context, each of the drawings included in the sample are supposed to be independent. In the presence of correlated data (such as correlated in time), again, other approaches, such as the one communicated in [41], are more suited.

4. Conclusions

A new test statistic to be used to measure the agreement between continuous theoretical distributions and samples drawn from TS was proposed. The analytical formula of the TS cumulative distribution function was obtained. The comparative study against other order statistics revealed that the newly proposed statistic carries distinct information regarding the quality of the sampling. A combined probability formula from a battery of statistics is suggested as a more accurate measure for the quality of the sampling. Therefore Equation (15) combining the probabilities (the risks of being in error) from Equation (5) to Equation (12) is recommended anytime when extreme values are suspected being outliers in samples from continuous distributions.

Supplementary Materials: The following are available online at http://www.mdpi.com/2227-7390/8/2/216/s1. The source code for sampling order statistics (file named OS.pas) and source code evaluation of the CDF of TS with Algorithm 4 (file named TS.pas file) are available upon request. The k-Sample Anderson–Darling test(s) on risks of being in error from the case studies 1 to 10 is given as a supplementary file.

Funding: This research received no external funding.

Acknowledgments: The following software were used during the research and writing the paper: Lazarus (freeware) were used to compile the 64bit executable for Monte Carlo sampling (using the parametrization given in Table 1). The executable was compiled to work for a 64GB multi-core workstation and were used so. Mathcad (v.14, licensed) were used to check the validity for some of the equations given (Equations (19)–(22), (24), (26), (27)), and to do the MLE estimates (implementing Equation (14) with first order derivatives and results given in Section 3.4 as Case studies 1 to 10). Matlab (v.8.5.0, licensed) was used to obtain Figures 4–8. Wolfram Mathematica (v.12.0, licensed) was used to check (iteratively) the formulas given for $1/TS$ (Equations (19) and 21)) and to provide the data for Figure 8. FreePascal (with GNU GMP, freeware) were used to assess numerically the agreement for TS statistic (Tables 2 and 3, Figure 8). StatSoft Statistica (v.7, licensed) was used to obtain Figures 11 and 12.

Conflicts of Interest: The author declares no conflict of interest.

References

1. Cramér, H. On the composition of elementary errors. *Scand. Actuar. J.* **1928**, *1*, 13–74. [CrossRef]
2. Von Mises, R.E. *Wahrscheinlichkeit, Statistik und Wahrheit*; Julius Springer: Berlin, Germany, 1928.
3. Watson, G.S. Goodness-of-fit tests on a circle. *Biometrika* **1961**, *48*, 109–114. [CrossRef]
4. Kolmogoroff, A. Sulla determinazione empirica di una legge di distribuzione. *Giornale dell'Istituto Italiano degli Attuari* **1933**, *4*, 83–91.
5. Kolmogoroff, A. Confidence limits for an unknown distribution function. *Ann. Math. Stat.* **1941**, *12*, 461–463. [CrossRef]
6. Smirnov, N. Table for estimating the goodness of fit of empirical distributions. *Ann. Math. Stat.* **1948**, *19*, 279–281. [CrossRef]
7. Kuiper, N.H. Tests concerning random points on a circle. *Proc. K. Ned. Akad. Wet. Ser. A* **1960**, *63*, 38–47. [CrossRef]
8. Anderson, T.W.; Darling, D. Asymptotic theory of certain 'goodness-of-fit' criteria based on stochastic processes. *Ann. Math. Stat.* **1952**, *23*, 193–212. [CrossRef]
9. Anderson, T.W.; Darling, D.A. A test of goodness of fit. *J. Am. Stat. Assoc.* **1954**, *49*, 765–769. [CrossRef]
10. Jäntschi, L.; Bolboacă, S.D. Performances of Shannon's entropy statistic in assessment of distribution of data. *Ovidius Univ. Ann. Chem.* **2017**, *28*, 30–42. [CrossRef]
11. Hilton, S.; Cairola, F.; Gardi, A.; Sabatini, R.; Pongsakornsathien, N.; Ezer, N. Uncertainty quantification for space situational awareness and traffic management. *Sensors* **2019**, *19*, 4361. [CrossRef]
12. Schöttl, J.; Seitz, M.J.; Köster, G. Investigating the randomness of passengers' seating behavior in suburban trains. *Entropy* **2019**, *21*, 600. [CrossRef]
13. Yang, X.; Wen, S.; Liu, Z.; Li, C.; Huang, C. Dynamic properties of foreign exchange complex network. *Mathematics* **2019**, *7*, 832. [CrossRef]
14. Młynski, D.; Bugajski, P.; Młynska, A. Application of the mathematical simulation methods for the assessment of the wastewater treatment plant operation work reliability. *Water* **2019**, *11*, 873. [CrossRef]
15. Jäntschi, L. A test detecting the outliers for continuous distributions based on the cumulative distribution function of the data being tested. *Symmetry* **2019**, *11*, 835. [CrossRef]
16. Metropolis, N.; Ulam, S. The Monte Carlo method. *J. Am. Stat. Assoc.* **1949**, *44*, 335–341. [CrossRef]
17. Jäntschi, L.; Bolboacă, S.D. Computation of probability associated with Anderson-Darling statistic. *Mathematics* **2018**, *6*, 88. [CrossRef]
18. Matsumoto, M.; Nishimura, T. Mersenne twister: A 623-dimensionally equidistributed uniform pseudo-random number generator. *ACM Trans. Model. Comput. Simul.* **1998**, *8*, 3–30. [CrossRef]
19. Fisher, R.A. On an absolute criterion for fitting frequency curves. *Messenger Math.* **1912**, *41*, 155–160,
20. Fisher, R.A. Questions and answers 14: Combining independent tests of significance. *Am. Stat.* **1948**, *2*, 30–31. [CrossRef]
21. Bolboacă, S.D.; Jäntschi, L.; Sestraş, A.F.; Sestraş, R.E.; Pamfil, D.C. Supplementary material of 'Pearson-Fisher chi-square statistic revisited'. *Information* **2011**, *2*, 528–545. [CrossRef]
22. Irwin, J.O. On the frequency distribution of the means of samples from a population having any law of frequency with finite moments, with special reference to Pearson's type II. *Biometrika* **1927**, *19*, 225–239. [CrossRef]
23. Hall, P. The distribution of means for samples of size N drawn from a population in which the variate takes values between 0 and 1, all such values being equally probable. *Biometrika* **1927**, *19*, 240–245. [CrossRef]
24. *Mathematica*, version 12.0; Software for Technical Computation; Wolfram Research: Champaign, IL, USA, 2019.
25. *GMP: The GNU Multiple Precision Arithmetic Library*, version 5.0.2; Software for Technical Computation; Free Software Foundation: Boston, MA, USA, 2016.

26. FreePascal: Open Source Compiler for Pascal and Object Pascal, Version 3.0.4. 2017. Available online: https://www.freepascal.org/ (accessed on 8 February 2020).
27. Pollet, T.V.; Meij, L. To remove or not to remove: the impact of outlier handling on significance testing in testosterone data. *Adapt. Hum. Behav. Physiol.* **2017**, *3*, 43–60. [CrossRef]
28. Wilks, S.S. Order statistics. *Bull. Am. Math. Soc.* **1948**, *54*, 6–50. [CrossRef]
29. Grubbs, F.E. Procedures for detecting outlying observations in samples. *Technometrics* **1969**, *11*, 1–21. [CrossRef]
30. Jäntschi, L.; Bolboacă, S.D. Distribution fitting 2. Pearson-Fisher, Kolmogorov-Smirnov, Anderson-Darling, Wilks-Shapiro, Kramer-von-Misses and Jarque-Bera statistics. *BUASVMCN Hortic.* **2009**, *66*, 691–697.
31. Bolboacă, S.D.; Jäntschi, L. Distribution fitting 3. Analysis under normality assumption. *BUASVMCN Hortic.* **2009**, *66*, 698–705.
32. Thomas, A.; Oommen, B.J. The fundamental theory of optimal 'Anti-Bayesian' parametric pattern classification using order statistics criteria. *Pattern Recognit.* **2013**, *46*, 376–388. [CrossRef]
33. Hu, L. A note on order statistics-based parametric pattern classification. *Pattern Recognit.* **2015**, *48*, 43–49. [CrossRef]
34. Jäntschi, L.; Bolboacă, S.D. Rarefaction on natural compound extracts diversity among genus. *J. Comput. Sci.* **2014**, *5*, 363–367. [CrossRef]
35. Jäntschi, L.; Bolboacă, S.D. Informational entropy of b-ary trees after a vertex cut. *Entropy* **2008**, *10*, 576–588. [CrossRef]
36. Scholz, F.W.; Stephens, M.A. K-sample Anderson-Darling tests. *J. Am. Stat. Assoc.* **1987**, *82*, 918–924. [CrossRef]
37. Xu, Z.; Huang, X.; Jimenez, F.; Deng, Y. A new record of graph enumeration enabled by parallel processing. *Mathematics* **2019**, *7*, 1214. [CrossRef]
38. Krizan, P.; Kozubek, M.; Lastovicka, J. Discontinuities in the ozone concentration time series from MERRA 2 reanalysis. *Atmosphere* **2019**, *10*, 812. [CrossRef]
39. Liang, K.; Zhang, Z.; Liu, P.; Wang, Z.; Jiang, S. Data-driven ohmic resistance estimation of battery packs for electric vehicles. *Energies* **2019**, *12*, 4772. [CrossRef]
40. Tamazian, A.; Nguyen, V.D.; Markelov, O.A.; Bogachev, M.I. Universal model for collective access patterns in the Internet traffic dynamics: A superstatistical approach. *EPL* **2016**, *115*, 10008. [CrossRef]
41. Nguyen, V.D.; Markelov, O.A.; Serdyuk, A.D.; Vasenev, A.N.; Bogachev, M.I. Universal rank-size statistics in network traffic: Modeling collective access patterns by Zipf's law with long-term correlations. *EPL* **2018**, *123*, 50001. [CrossRef]

© 2020 by the author. Licensee MDPI, Basel, Switzerland. This article is an open access article distributed under the terms and conditions of the Creative Commons Attribution (CC BY) license (http://creativecommons.org/licenses/by/4.0/).

Article

dCATCH—A Numerical Package for d-Variate near G-Optimal Tchakaloff Regression via Fast NNLS

Monica Dessole, Fabio Marcuzzi and Marco Vianello *

Department of Mathematics "Tullio Levi Civita", University of Padova, Via Trieste 63, 35131 Padova, Italy; mdessole@math.unipd.it (M.D.); marcuzzi@math.unipd.it (F.M.)
* Correspondence: marcov@math.unipd.it

Received: 11 June 2020; Accepted: 7 July 2020; Published: 9 July 2020

Abstract: We provide a numerical package for the computation of a d-variate near G-optimal polynomial regression design of degree m on a finite design space $X \subset \mathbb{R}^d$, by few iterations of a basic multiplicative algorithm followed by Tchakaloff-like compression of the discrete measure keeping the reached G-efficiency, via an accelerated version of the Lawson-Hanson algorithm for Non-Negative Least Squares (NNLS) problems. This package can solve on a personal computer large-scale problems where $card(X) \times \dim(P_{2m}^d)$ is up to 10^8–10^9, being $\dim(P_{2m}^d) = \binom{2m+d}{d} = \binom{2m+d}{2m}$. Several numerical tests are presented on complex shapes in $d=3$ and on hypercubes in $d>3$.

Keywords: multivariate polynomial regression designs; G-optimality; D-optimality; multiplicative algorithms; G-efficiency; Caratheodory-Tchakaloff discrete measure compression; Non-Negative Least Squares; accelerated Lawson-Hanson solver

1. Introduction

In this paper we present the numerical software package *dCATCH* [1] for the computation of a d-variate near G-optimal polynomial regression design of degree m on a finite design space $X \subset \mathbb{R}^d$. In particular, it is the first software package for general-purpose Tchakaloff-like compression of d-variate designs via Non-Negative Least Squares (NNLS), freely available on the Internet. The code is an evolution of the codes in Reference [2] (limited to $d=2,3$), with a number of features tailored to higher dimension and large-scale computations. The key ingredients are:

- use of d-variate Vandermonde-like matrices at X in a discrete orthogonal polynomial basis (obtained by discrete orthonormalization of the total-degree product Chebyshev basis of the minimal box containing X), with automatic adaptation to the actual dimension of $\mathbb{P}_m^d(X)$;
- few tens of iterations of the basic Titterington multiplicative algorithm until near G-optimality of the design is reached, with a checked G-efficiency of say 95% (but with a design support still far from sparsity);
- Tchakaloff-like compression of the resulting near G-optimal design via NNLS solution of the underdetermined moment system, with concentration of the discrete probability measure by sparse re-weighting to a support $\subset X$, of cardinality at most $\mathbb{P}_{2m}^d(X)$, keeping the same G-efficiency;
- iterative solution of the large-scale NNLS problem by a new accelerated version of the classical Lawson-Hanson active set algorithm, that we recently introduced in Reference [3] for 2d and 3d instances and here we validate on higher dimensions.

Before giving a more detailed description of the algorithm, it is worth recalling in brief some basic notions of optimal design theory. Such a theory has its roots and main applications within statistics, but also strong connections with approximation theory. In statistics, a design is a probability measure μ supported on a (discrete or continuous) compact set $\Omega \subset \mathbb{R}^d$. The search for designs that optimize some properties of statistical estimators (optimal designs) dates back to at least one century ago, and the relevant literature is so wide and still actively growing and monographs and survey papers are abundant in the literature. For readers interested in the evolution and state of the art of this research field, we may quote, for example, two classical treatises such as in References [4,5], the recent monograph [6] and the algorithmic survey [7], as well as References [8–10] and references therein. On the approximation theory side we may quote, for example, References [11,12].

The present paper is organized as follows—in Section 2 we briefly recall some basic concepts from the theory of Optimal Designs, for the reader's convenience, with special attention to the deterministic and approximation theoretic aspects. In Section 3 we present in detail our computational approach to near G-optimal d-variate designs via Caratheodory-Tchakaloff compression. All the routines of the *dCATCH* software package here presented, are described. In Section 4 we show several numerical results with dimensions in the range 3–10 and a Conclusions section follows.

For the reader's convenience we also display Tables 1 and 2, describing the acronyms used in this paper and the content (subroutine names) of the *dCATCH* software package.

Table 1. List of acronyms.

LS	Least Squares
NNLS	Non-Negative Least Squares
LH	Lawson-Hawson algorithm for NNLS
LHI	Lawson-Hawson algorithm with unconstrained LS Initialization
LHDM	Lawson-Hawson algorithm with Deviation Maximization acceleration

Table 2. *dCATCH* package content.

dCATCH	d-variate CAratheodory-TCHakaloff discrete measure compression
dCHEBVAND	d-variate Chebyshev-Vandermonde matrix
dORTHVAND	d-variate Vandermonde-like matrix in a weighted orthogonal polynomial basis
dNORD	d-variate Near G-Optimal Regression Designs
LHDM	Lawson-Hawson algorithm with Deviation Maximization acceleration

2. G-Optimal Designs

Let $\mathbb{P}_m^d(\Omega)$ denote the space of d-variate real polynomials of total degree not greater than n, restricted to a (discrete or continuous) compact set $\Omega \subset \mathbb{R}^d$, and let μ be a design, that is, a probability measure, with $supp(\mu) \subseteq \Omega$. In what follows we assume that $supp(\mu)$ is *determining* for $\mathbb{P}_m^d(\Omega)$ [13], that is, polynomials in $\mathbb{P}_m^d$ vanishing on $supp(\mu)$ vanish everywhere on Ω.

In the theory of optimal designs, a key role is played by the diagonal of the reproducing kernel for μ in $\mathbb{P}_m^d(\Omega)$ (also called the Christoffel polynomial of degree m for μ)

$$K_m^\mu(x,x) = \sum_{j=1}^{N_m} p_j^2(x), \quad N_m = \dim(\mathbb{P}_m^d(\Omega)), \tag{1}$$

where $\{p_j\}$ is any μ-orthonormal basis of $\mathbb{P}_m^d(\Omega)$. Recall that $K_m^\mu(x,x)$ can be proved to be independent of the choice of the orthonormal basis. Indeed, a relevant property is the following estimate of the L^∞-norm in terms of the L_μ^2-norm of polynomials

$$\|p\|_{L^\infty(\Omega)} \leq \sqrt{\max_{x \in \Omega} K_m^\mu(x,x)} \, \|p\|_{L_\mu^2(\Omega)}, \quad \forall p \in \mathbb{P}_m^d(\Omega). \tag{2}$$

Now, by (1) and μ-orthonormality of the basis we get

$$\int_\Omega K_m^\mu(x,x) \, d\mu = \sum_{j=1}^{N_m} \int_\Omega p_j^2(x) \, d\mu = N_m, \tag{3}$$

which entails that $\max_{x\in\Omega} K_m^\mu(x,x) \geq N_m$.

Then, a probability measure $\mu_* = \mu_*(\Omega)$ is then called a G-optimal design for polynomial regression of degree m on Ω if

$$\min_\mu \max_{x\in\Omega} K_m^\mu(x,x) = \max_{x\in\Omega} K_m^{\mu_*}(x,x) = N_m . \qquad (4)$$

Observe that, since $\int_\Omega K_m^\mu(x,x)\,d\mu = N_m$ for every μ, an optimal design has also the following property $K_m^{\mu_*}(x,x) = N_m$, μ_*-a.e. in Ω.

Now, the well-known Kiefer-Wolfowitz General Equivalence Theorem [14] (a cornerstone of optimal design theory), asserts that the difficult min-max problem (4) is equivalent to the much simpler maximization problem

$$\max_\mu \det(G_m^\mu), \quad G_m^\mu = \left(\int_\Omega \phi_i(x)\phi_j(x)\,d\mu\right)_{1\leq i,j\leq N_m},$$

where G_m^μ is the Gram matrix (or information matrix in statistics) of μ in a fixed polynomial basis $\{\phi_i\}$ of $\mathbb{P}_m^d(\Omega)$. Such an optimality is called D-optimality, and ensures that an optimal measure always exists, since the set of Gram matrices of probability measures is compact and convex; see for example, References [5,12] for a general proof of these results, valid for continuous as well as for discrete compact sets.

Notice that an optimal measure is neither unique nor necessarily discrete (unless Ω is discrete itself). Nevertheless, the celebrated Tchakaloff Theorem ensures the existence of a positive quadrature formula for integration in $d\mu_*$ on Ω, with cardinality not exceeding $N_{2m} = \dim(\mathbb{P}_{2m}^d(\Omega))$ and which is exact for all polynomials in $\mathbb{P}_{2m}^d(\Omega)$. Such a formula is then a design itself, and it generates the same orthogonal polynomials and hence the same Christoffel polynomial of μ_*, preserving G-optimality (see Reference [15] for a proof of Tchakaloff Theorem with general measures).

We recall that G-optimality has two important interpretations in terms of statistical and deterministic polynomial regression.

From a statistical viewpoint, it is the probability measure on Ω that minimizes the maximum prediction variance by polynomial regression of degree m, cf. for example, Reference [5].

On the other hand, from an approximation theory viewpoint, if we call $\mathcal{L}_m^{\mu_*}$ the corresponding weighted least squares projection operator $L^\infty(\Omega) \to \mathbb{P}_m^d(\Omega)$, namely

$$\|f - \mathcal{L}_m^{\mu_*} f\|_{L^2_{\mu_*}(\Omega)} = \min_{p\in\mathbb{P}_m^d(\Omega)} \|f - p\|_{L^2_{\mu_*}(\Omega)}, \qquad (5)$$

by (2) we can write for every $f \in L^\infty(\Omega)$

$$\|\mathcal{L}_m^{\mu_*} f\|_{L^\infty(\Omega)} \leq \sqrt{\max_{x\in\Omega} K_m^{\mu_*}(x,x)}\,\|\mathcal{L}_m^{\mu_*} f\|_{L^2_{\mu_*}(\Omega)} = \sqrt{N_m}\,\|\mathcal{L}_m^{\mu_*} f\|_{L^2_{\mu_*}(\Omega)}$$

$$\leq \sqrt{N_m}\,\|f\|_{L^2_{\mu_*}(\Omega)} \leq \sqrt{N_m}\,\|f\|_{L^\infty(\Omega)},$$

(where the second inequality comes from μ_*-orthogonality of the projection), which gives

$$\|\mathcal{L}_m^{\mu_*}\| = \sup_{f\neq 0} \frac{\|\mathcal{L}_m^{\mu_*} f\|_{L^\infty(\Omega)}}{\|f\|_{L^\infty(\Omega)}} \leq \sqrt{N_m}, \qquad (6)$$

that is a G-optimal measure minimizes (the estimate of) the weighted least squares uniform operator norm.

We stress that in this paper we are interested in the fully discrete case of a finite design space $\Omega = X$, so that any design μ is identified by a set of positive weights (masses) summing up to 1 and integrals are weighted sums.

3. Computing near G-Optimal Compressed Designs

Since in the present context we have a finite design space $\Omega = X = \{x_1, \ldots, x_M\} \subset \mathbb{R}^d$, we may think a design μ as a vector of non-negative weights $u = (u_1, \cdots, u_M)$ attached to the points, such that $\|u\|_1 = 1$ (the support of μ being identified by the positive weights). Then, a G-optimal (or D-optimal) design μ_* is represented by the corresponding non-negative vector u_*. We write $K_m^u(x, x) = K_m^\mu(x, x)$ for the Christoffel polynomial and similarly for other objects (spaces, operators, matrices) corresponding to a discrete design. At the same time, $L^\infty(\Omega) = \ell^\infty(X)$, and $L_\mu^2(\Omega) = \ell_u^2(X)$ (a weighted ℓ^2 functional space on X) with $\|f\|_{\ell_u^2(X)} = \left(\sum_{i=1}^M u_i f^2(x_i)\right)^{1/2}$.

In order to compute an approximation of the desired u_*, we resort to the basic multiplicative algorithm proposed by Titterington in the '70s (cf. Reference [16]), namely

$$u_i(k+1) = u_i(k) \frac{K_m^{u(k)}(x_i, x_i)}{N_m}, \quad 1 \leq i \leq M, \quad k = 0, 1, 2, \ldots, \qquad (7)$$

with initialization $u(0) = (1/M, \ldots, 1/M)^T$. Such an algorithm is known to be convergent sublinearly to a D-optimal (or G-optimal by the Kiefer-Wolfowitz Equivalence Theorem) design, with an increasing sequence of Gram determinants

$$det(G_m^{u(k)}) = det(V^T diag(u(k)) V),$$

where V is a Vandermonde-like matrix in any fixed polynomial basis of $\mathbb{P}_m^d(X)$; cf., for example, References [7,10]. Observe that $u(k+1)$ is indeed a vector of positive probability weights if such is $u(k)$. In fact, the Christoffel polynomial $K_m^{u(k)}$ is positive on X, and calling μ_k the probability measure on X associated with the weights $u(k)$ we get immediately $\sum_i u_i(k+1) = \frac{1}{N_m} \sum_i u_i(k) K_m^{u(k)}(x_i, x_i) = \frac{1}{N_m} \int_X K_m^{u(k)}(x, x) d\mu_k = 1$ by (3) in the discrete case $\Omega = X$.

Our implementation of (7) is based on the functions

- $C = $ dCHEBVAND(n, X)
- $[U, jvec] = $ dORTHVAND$(n, X, u, jvec)$
- $[pts, w] = $ dNORD$(m, X, gtol)$

The function dCHEBVAND computes the d-variate Chebyshev-Vandermonde matrix $C = (\phi_j(x_i)) \in \mathbb{R}^{M \times N_n}$, where $\{\phi_j(x)\} = \{T_{v_1}(\alpha_1 x_1 + \beta_1) \ldots T_{v_d}(\alpha_d x_d + \beta_d)\}, 0 \leq v_i \leq n, v_1 + \cdots + v_d \leq n$, is a suitably ordered total-degree product Chebyshev basis of the minimal box $[a_1, b_1] \times \cdots \times [a_d, b_d]$ containing X, with $\alpha_i = 2/(b_i - a_i), \beta_i = -(b_i + a_i)/(b_i - a_i)$. Here we have resorted to the codes in Reference [17] for the construction and enumeration of the required "monomial" degrees. Though the initial basis is then orthogonalized, the choice of the Chebyshev basis is dictated by the necessity of controlling the conditioning of the matrix. This would be on the contrary extremely large with the standard monomial basis, already at moderate regression degrees, preventing a successful orthogonalization.

Indeed, the second function dORTHVAND computes a Vandermonde-like matrix in a u-orthogonal polynomial basis on X, where u is the probability weight array. This is accomplished essentially by numerical rank evaluation for $C = $ dCHEBVAND(n, X) and QR factorization

$$diag(\sqrt{u}) C_0 = QR, \quad U = C_0 R^{-1}, \qquad (8)$$

(with Q orthogonal rectangular and R square invertible), where $\sqrt{u} = (\sqrt{u_1}, \ldots, \sqrt{u_M})$. The matrix C_0 has full rank and corresponds to a selection of the columns of C (i.e., of the original basis polynomials) via QR with column pivoting, in such a way that these form a basis of $\mathbb{P}_n^d(X)$, since $rank(C) = dim(\mathbb{P}_n^d(X))$. A possible alternative, not yet implemented, is the direct use of a rank-revealing QR factorization. The in-out parameter "jvec" allows to pass directly the column index vector corresponding to a polynomial basis after a previous call to dORTHVAND with the same degree n, avoiding numerical rank computation and allowing a simple "economy size" QR factorization of $diag(\sqrt{u})\, C_0 = diag(\sqrt{u})\, C(:,jvec)$.

Summarizing, U is a Vandermonde-like matrix for degree n on X in the required u-orthogonal basis of $\mathbb{P}_n^d(X)$, that is

$$[p_1(x),\ldots,p_{N_n}(x)] = [\phi_{j_1}(x),\ldots,\phi_{j_{N_n}}(x)]\, R^{-1}, \tag{9}$$

where $jvec = (j_1,\ldots,j_{N_n})$ is the multi-index resulting from pivoting. Indeed by (8) we can write the scalar product $(p_h,p_k)_{\ell_u^2(X)}$ as

$$(p_h,p_k)_{\ell_u^2(X)} = \sum_{i=1}^M u_i\, p_h(x_i)\, p_k(x_i) = (U^T diag(u)\, U)_{hk} = (Q^T Q)_{hk} = \delta_{hk},$$

for $1 \leq h,k \leq N_n$, which shows orthonormality of the polynomial basis in (9).

We stress that $rank(C) = \dim(\mathbb{P}_n^d(X))$ could be strictly smaller than $\dim(\mathbb{P}_n^d) = \binom{n+d}{d}$, when there are polynomials in $\mathbb{P}_n^d$ vanishing on X that do not vanish everywhere. In other words, X lies on a lower-dimensional algebraic variety (technically one says that X is not $\mathbb{P}_n^d$-determining [13]). This certainly happens when $card(X)$ is too small, namely $card(X) < \dim(\mathbb{P}_n^d)$, but think for example also to the case when $d = 3$ and X lies on the 2-sphere S^2 (independently of its cardinality), then we have $\dim(\mathbb{P}_n^d(X)) \leq \dim(\mathbb{P}_n^d(S^2)) = (n+1)^2 < \dim(\mathbb{P}_n^3) = (n+1)(n+2)(n+3)/6$.

Iteration (7) is implemented within the third function dNORD whose name stands for d-dimensional Near G-Optimal Regression Designs, which calls dORTHVAND with $n = m$. Near optimality is here twofold, namely it concerns both the concept of G-efficiency of the design and the sparsity of the design support.

We recall that G-efficiency is the percentage of G-optimality reached by a (discrete) design, measured by the ratio

$$G_m(u) = \frac{N_m}{max_{x \in X} K_m^u(x,x)},$$

knowing that $G_m(u) \leq 1$ by (3) in the discrete case $\Omega = X$. Notice that $G_m(u)$ can be easily computed after the construction of the u-orthogonal Vandermonde-like matrix U by dORTHVAND, as $G_m(u) = N_m / (\max_i \|row_i(U)\|_2^2)$.

In the multiplicative algorithm (7), we then stop iterating when a given threshold of G-efficiency (the input parameter "gtol" in the call to dNORD) is reached by $u(k)$, since $G_m(u(k)) \to 1$ as $k \to \infty$, say for example $G_m(u(k)) \geq 95\%$ or $G_m(u(k)) \geq 99\%$. Since convergence is sublinear and in practice we see that $1 - G_m(u(k)) = \mathcal{O}(1/k)$, for a 90% G-efficiency the number of iterations is typically in the tens, whereas it is in the hundreds for 99% one and in the thousands for 99,9%. When a G-efficiency very close to 1 is needed, one could resort to more sophisticated multiplicative algorithms, see for example, References [9,10].

In many applications however a G-efficiency of 90–95% could be sufficient (then we may speak of near G-optimality of the design), but though in principle the multiplicative algorithm converges to an optimal design μ_* on X with weights u_* and cardinality $N_m \leq card(supp(\mu_*)) \leq N_{2m}$, such a sparsity is far from being reached after the iterations that guarantee near G-optimality, in the sense that there is a still large percentage of non-negligible weights in the near optimal design weight vector, say

$$u(\bar{k}) \text{ such that } G_m(u(\bar{k})) \geq gtol. \tag{10}$$

Following References [18,19], we can however effectively compute a design which has the same G-efficiency of $u(\bar{k})$ but a support with a cardinality not exceeding $N_{2m} = \dim(\mathbb{P}_{2m}^d(X))$, where in many applications $N_{2m} \ll card(X)$, obtaining a remarkable compression of the near optimal design.

The theoretical foundation is a generalized version [15] of Tchakaloff Theorem [20] on positive quadratures, which asserts that for every measure on a compact set $\Omega \subset \mathbb{R}^d$ there exists an algebraic quadrature formula exact on $\mathbb{P}_n^d(\Omega)$), with positive weights, nodes in Ω and cardinality not exceeding $N_n = \dim(\mathbb{P}_n^d(\Omega))$.

In the present discrete case, that is, where the designs are defined on $\Omega = X$, this theorem implies that for every design μ on X there exists a design ν, whose support is a subset of X, which is exact for integration in $d\mu$ on $\mathbb{P}_n^d(X)$. In other words, the design ν has the same basis moments (indeed, for any basis of $\mathbb{P}_n^d(\Omega)$)

$$\int_X p_j(x)\, d\mu = \sum_{i=1}^M u_i\, p_j(x_i) = \int_X p_j(x)\, d\nu = \sum_{\ell=1}^L w_\ell\, p_j(\xi_\ell)\,, \quad 1 \leq j \leq N_n\,,$$

where $L \leq N_n \leq M$, $\{u_i\}$ are the weights of μ, $supp(\nu) = \{\xi_\ell\} \subseteq X$ and $\{w_\ell\}$ are the positive weights of ν. For $L < M$, which certainly holds if $N_n < M$, this represents a compression of the design μ into the design ν, which is particularly useful when $N_n \ll M$.

In matrix terms this can be seen as the fact that the underdetermined $\{p_j\}$-moment system

$$U_n^T v = U_n^T u \tag{11}$$

has a non-negative solution $v = (v_1, \ldots, v_M)^T$ whose positive components, say $w_\ell = v_{i_\ell}, 1 \leq \ell \leq L \leq N_n$, determine the support points $\{\xi_\ell\} \subseteq X$ (for clarity we indicate here by U_n the matrix U computed by dORTHVAND at degree n). This fact is indeed a consequence of the celebrated Caratheodory Theorem on conic combinations [21], asserting that a linear combination with non-negative coefficients of M vectors in $\mathbb{R}^N$ with $M > N$ can be re-written as linear positive combination of at most N of them. So, we get the discrete version of Tchakaloff Theorem by applying Caratheodory Theorem to the columns of U_n^T in the system (11), ensuring then existence of a non-negative solution v with at most N_n nonzero components.

In order to compute such a solution to (11) we choose the strategy based on Quadratic Programming introduced in Reference [22], namely on sparse solution of the Non-Negative Least Squares (NNLS) problem

$$v = \mathrm{argmin}_{z \in \mathbb{R}^M, z \geq 0} \|U_n^T z - U_n^T u\|_2^2$$

by a new accelerated version of the classical Lawson-Hanson active-set method, proposed in Reference [3] in the framework of design optimization in $d = 2, 3$ and implemented by the function LHDM (Lawson-Hanson with Deviation Maximization), that we tune in the present package for very large-scale d-variate problems (see the next subsection for a brief description and discussion). We observe that working with an orthogonal polynomial basis of $\mathbb{P}_n^d(X)$ allows to deal with the well-conditioned matrix U_n in the Lawson-Hanson algorithm.

The overall computational procedure is implemented by the function

- $[pts, w, momerr] = \mathrm{dCATCH}(n, X, u)$,

where dCATCH stands for d-variate CAratheodory-TCHakaloff discrete measure compression. It works for any discrete measure on a discrete set X. Indeed, it could be used, other than for design compression, also in the compression of d-variate quadrature formulas, to give an example. The output parameter $pts = \{\xi_\ell\} \subset X$ is the array of support points of the compressed measure, while $w = \{w_\ell\} = \{v_{i_\ell} > 0\}$ is the corresponding positive weight array (that we may call a d-variate near G-optimal Tchakaloff design) and $momerr = \|U_n^T v - U_n^T u\|_2$ is the moment residual. This function is called LHDM.

In the present framework we call dCATCH with $n = 2m$ and $u = u(\bar{k})$, cf. (10), that is, we solve

$$v = \mathrm{argmin}_{z \in \mathbb{R}^M, z \geq 0} \|U_{2m}^T z - U_{2m}^T u(\bar{k})\|_2^2\,. \tag{12}$$

In such a way the compressed design generates the same scalar product of $u(\bar{k})$ in $\mathbb{P}_m^d(X)$, and hence the same orthogonal polynomials and the same Christoffel function on X keeping thus invariant the G-efficiency

$$\mathbb{P}_{2m}^d(X) \ni K_m^v(x,x) = K_m^{u(\bar{k})}(x,x) \; \forall x \in X \implies G_m(v) = G_m(u(\bar{k})) \geq gtol \tag{13}$$

with a (much) smaller support.

From a deterministic regression viewpoint (approximation theory), let us denote by p_m^{opt} the polynomial in $\mathbb{P}_m^d(X)$ of best uniform approximation for f on X, where we assume $f \in C(D)$ with $X \subset D \subset \mathbb{R}^d$, D being a compact domain (or even lower-dimensional manifold), and by $E_m(f;X) = \inf_{p \in \mathbb{P}_m^d(X)} \|f - p\|_{\ell^\infty(X)} = \|f - p_m^{opt}\|_{\ell^\infty(X)}$ and $E_m(f;D) = \inf_{p \in \mathbb{P}_m^d(D)} \|f - p\|_{L^\infty(D)}$ the best uniform polynomial approximation errors on X and D.

Then, denoting by $\mathcal{L}_m^{u(\bar{k})}$ and $\mathcal{L}_m^w f = \mathcal{L}_m^v f$ the weighted least squares polynomial approximation of f (cf. (5)) by the near G-optimal weights $u(\bar{k})$ and w, respectively, with the same reasoning used to obtain (6) and by (13) we can write the operator norm estimates

$$\|\mathcal{L}_m^{u(\bar{k})}\|, \|\mathcal{L}_m^w\| \leq \sqrt{\tilde{N}_m} \leq \sqrt{\frac{N_m}{gtol}}, \quad \tilde{N}_m = \frac{N_m}{G_m(u(\bar{k}))} = \frac{N_m}{G_m(v)}.$$

Moreover, since $\mathcal{L}_m^w p = p$ for any $p \in \mathbb{P}_m^d(X)$, we can write the near optimal estimate

$$\|f - \mathcal{L}_m^w f\|_{\ell^\infty(X)} \leq \|f - p_m^{opt}\|_{\ell^\infty(X)} + \|p_m^{opt} - \mathcal{L}_m^w p_m^{opt}\|_{\ell^\infty(X)} + \|\mathcal{L}_m^w p_m^{opt} - \mathcal{L}_m^w f\|_{\ell^\infty(X)}$$

$$= \|f - p_m^{opt}\|_{\ell^\infty(X)} + \|\mathcal{L}_m^w p_m^{opt} - \mathcal{L}_m^w f\|_{\ell^\infty(X)} \leq (1 + \|\mathcal{L}_m^w\|) E_m(f;X)$$

$$\leq \left(1 + \sqrt{\frac{N_m}{gtol}}\right) E_m(f;X) \leq \left(1 + \sqrt{\frac{N_m}{gtol}}\right) E_m(f;D) \approx \left(1 + \sqrt{N_m}\right) E_m(f;D).$$

Notice that $\mathcal{L}_m^w f$ is constructed by sampling f only at the compressed support $\{\xi_\ell\} \subset X$. The error depends on the regularity of f on $D \supset X$, with a rate that can be estimated whenever D admits a multivariate Jackson-like inequality, cf. Reference [23].

Accelerating the Lawson-Hanson Algorithm by Deviation Maximization (LHDM)

Let $A \in \mathbb{R}^{N \times M}$ and $b \in \mathbb{R}^N$. The NNLS problem consists of seeking $x \in \mathbb{R}^M$ that solves

$$x = \mathrm{argmin}_{z \geq 0} \|Az - b\|_2^2. \tag{14}$$

This is a convex optimization problem with linear inequality constraints that define the *feasible region*, that is the positive orthant $\{x \in \mathbb{R}^M : x_i \geq 0\}$. The very first algorithm dedicated to problem (14) is due to Lawson and Hanson [24] and it is still one of the most often used. It was originally derived for solving overdetermined linear systems, with $N \gg M$. However, in the case of underdetermined linear systems, with $N \ll M$, this method succeeds in sparse recovery.

Recall that for a given point x in the feasible region, the index set $\{1, \ldots, M\}$ can be partitioned into two sets: the active set Z, containing the indices of active constraints $x_i = 0$, and the passive set P, containing the remaining indices of inactive constraints $x_i > 0$. Observe that an optimal solution x^* of (14) satisfies $Ax^* = b$ and, if we denote by P^* and Z^* the corresponding passive and active sets respectively, x^* also solves in a least square sense the following unconstrained least squares subproblem

$$x_{P^*}^* = \mathrm{argmin}_y \|A_{P^*} y - b\|_2^2, \tag{15}$$

where $A_{P^\star}$ is the submatrix containing the columns of A with index in $P^\star$, and similarly $x_{P^\star}^\star$ is the subvector made of the entries of $x^\star$ whose index is in $P^\star$. The remaining entries of $x^\star$, namely those whose index is in $Z^\star$, are null.

The Lawson-Hanson algorithm, starting from a null initial guess $x = 0$ (which is feasible), incrementally builds an optimal solution by moving indices from the active set Z to the passive set P and vice versa, while keeping the iterates within the feasible region. More precisely, at each iteration first order information is used to detect a column of the matrix A such that the corresponding entry in the new solution vector will be strictly positive; the index of such a column is moved from the active set Z to the passive set P. Since there's no guarantee that the other entries corresponding to indices in the former passive set will stay positive, an inner loop ensures the new solution vector falls into the feasible region, by moving from the passive set P to the active set Z all those indices corresponding to violated constraints. At each iteration a new iterate is computed by solving a least squares problem of type (15): this can be done, for example, by computing a QR decomposition, which is substantially expensive. The algorithm terminates in a finite number of steps, since the possible combinations of passive/active set are finite and the sequence of objective function values is strictly decreasing, cf. Reference [24].

The *deviation maximization* (DM) technique is based on the idea of adding a whole set of indices T to the passive set at each outer iteration of the Lawson-Hanson algorithm. This corresponds to select a block of new columns to insert in the matrix A_P, while keeping the current solution vector within the feasible region in such a way that sparse recovery is possible when dealing with non-strictly convex problems. In this way, the number of total iterations and the resulting computational cost decrease. The set T is initialized to the index chosen by the standard Lawson-Hanson (LH) algorithm, and it is then extended, within the same iteration, using a set of candidate indices C chosen is such a way that the corresponding entries are likely positive in the new iterate. The elements of T are then chosen carefully within C: note that if the columns corresponding to the chosen indices are linearly dependent, the submatrix of the least squares problem (15) will be rank deficient, leading to numerical difficulties. We add k new indices, where k is an integer parameter to tune on the problem size, in such a way that, at the end, for every pair of indices in the set T, the corresponding column vectors form an angle whose cosine in absolute value is below a given threshold *thres*. The whole procedure is implemented in the function

- $[x, resnorm, exitflag] = \text{LHDM}(A, b, options)$.

The input variable *options* is a structure containing the user parameters for the LHDM algorithm; for example, the aforementioned k and *thres*. The output parameter x is the least squares solution, *resnorm* is the squared 2-norm of the residual and *exitflag* is set to 0 if the LHDM algorithm has reached the maximum number of iterations without converging and 1 otherwise.

In the literature, an accelerating technique was introduced by Van Benthem and Keenan [25], who presented a different NNLS solution algorithm, namely "fast combinatorial NNLS", designed for the specific case of a large number of right-hand sides. The authors exploited a clever reorganization of computations in order to take advantage of the combinatorial nature of the problems treated (multivariate curve resolution) and introduced a nontrivial initialization of the algorithm by means of unconstrained least squares solution. In the following section we are going to compare such an approach, briefly named LHI, and the standard LH algorithm with the LHDM procedure just summarized.

4. Numerical Examples

In this section, we perform several tests on the computation of d-variate near G-optimal Tchakaloff designs, from low to moderate dimension d. In practice, we are able to treat, on a personal computer, large-scale problems where $card(X) \times \dim(P_{2m}^d)$ is up to 10^8–10^9, with $\dim(P_{2m}^d) = \binom{2m+d}{d} = \binom{2m+d}{2m}$.

Recall that the main memory requirement is given by the $N_{2m} \times M$ matrix U^T in the compression process solved by the LHDM algorithm, where $M = card(X)$ and $N_{2m} = \dim(P_{2m}^d(X)) \leq \dim(P_{2m}^d)$.

Given the dimension $d > 1$ and the polynomial degree m, the routine LHDM empirically sets the parameter k as follows $k = \lceil \binom{2m+d}{d}/(m(d-1)) \rceil$, while the threshold is $thres = \cos(\frac{\pi}{2} - \theta), \theta \approx 0.22$. All the tests are performed on a workstation with a 32 GB RAM and an Intel Core i7-8700 CPU @ 3.20 GHz.

4.1. Complex 3d Shapes

To show the flexibility of the package dCATCH, we compute near G-optimal designs on a "multibubble" $D \subset \mathbb{R}^3$ (i.e., the union of a finite number of non-disjoint balls), which can have a very complex shape with a boundary surface very difficult to describe analytically. Indeed, we are able to implement near optimal regression on quite complex solids, arising from finite union, intersection and set difference of simpler pieces, possibly multiply-connected, where for each piece we have available the indicator function via inequalities. Grid-points or low-discrepancy points, for example, Halton points, of a surrounding box, could be conveniently used to discretize the solid. Similarly, thanks to the adaptation of the method to the actual dimension of the polynomial spaces, we can treat near optimal regression on the surfaces of such complex solids, as soon as we are able to discretize the surface of each piece by point sets with good covering properties (for example, we could work on the surface of a multibubble by discretizing each sphere via one of the popular spherical point configurations, cf. Reference [26]).

We perform a test at regression degree $m = 10$ on the 5-bubble shown in Figure 1b. The initial support X consists in the $M = 18,915$ points within 64,000 low discrepancy Halton points, falling in the closure of the multibubble. Results are shown in Figure 1 and Table 3.

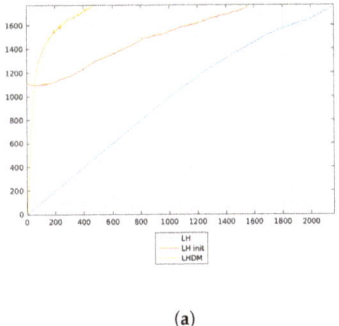

(a)

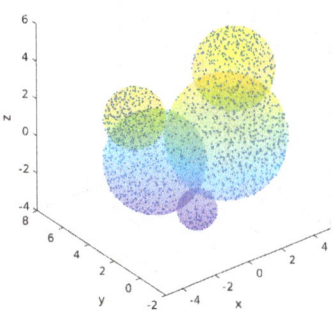

(b)

Figure 1. Multibubble test case, regression degree $m = 10$. (a) The evolution of the cardinality of the passive set P along the iterations of the three LH algorithms. (b) Multibubble with 1763 compressed Tchakaloff points, extracted from 18,915 original points.

Table 3. Results for the multibubble numerical test: $compr = M/mean(cpts)$ is the mean compression ratio obtained by the three methods listed; t_{LH}/t_{Titt} is the ratio between the execution time of LH and that of the Titterington algorithm; t_{LH}/t_{LHDM} (t_{LHI}/t_{LHDM}) is the ratio between the execution time of LH (LHI) and that of LHDM; $cpts$ is the number of compressed Tchakaloff points and $momerr$ is the final moment residual.

Test				LH			LHI			LHDM	
m	M	compr	t_{LH}/t_{Titt}	t_{LH}/t_{LHDM}	cpts	momerr	t_{LHI}/t_{LHDM}	cpts	momerr	cpts	momerr
10	18,915	11/1	40.0/1	2.7/1	1755	3.4×10^{-8}	3.2/1	1758	3.2×10^{-8}	1755	1.5×10^{-8}

4.2. Hypercubes: Chebyshev Grids

In a recent paper [19], a connection has been studied between the statistical notion of G-optimal design and the approximation theoretic notion of admissible mesh for multivariate polynomial approximation, deeply studied in the last decade after Reference [13] (see, e.g., References [27,28] with the references therein). In particular, it has been shown that near G-optimal designs on admissible meshes of suitable cardinality have a G-efficiency on the whole d-cube that can be made convergent to 1. For example, it has been proved by the notion of Dubiner distance and suitable multivariate polynomial inequalities, that a design with G-efficiency γ on a grid X of $(2km)^d$ Chebyshev points (the zeros of $T_{2km}(t) = \cos(2km \arccos(t))$, $t \in [-1,1]$), is a design for $[-1,1]^d$ with G-efficiency $\gamma(1 - \pi^2/(8k^2))$. For example, taking $k = 3$ a near G-optimal Tchakaloff design with $\gamma = 0.99$ on a Chebyshev grid of $(6m)^d$ points is near G-optimal on $[-1,1]^d$ with G-efficiency approximately $0.99 \cdot 0.86 \approx 0.85$, and taking $k = 4$ (i.e., a Chebyshev grid of $(8m)^d$ points) the corresponding G-optimal Tchakaloff design has G-efficiency approximately $0.99 \cdot 0.92 \approx 0.91$ on $[-1,1]^d$ (in any dimension d).

We perform three tests in different dimension spaces and at different regression degrees. Results are shown in Figure 2 and Table 4, using the same notation above.

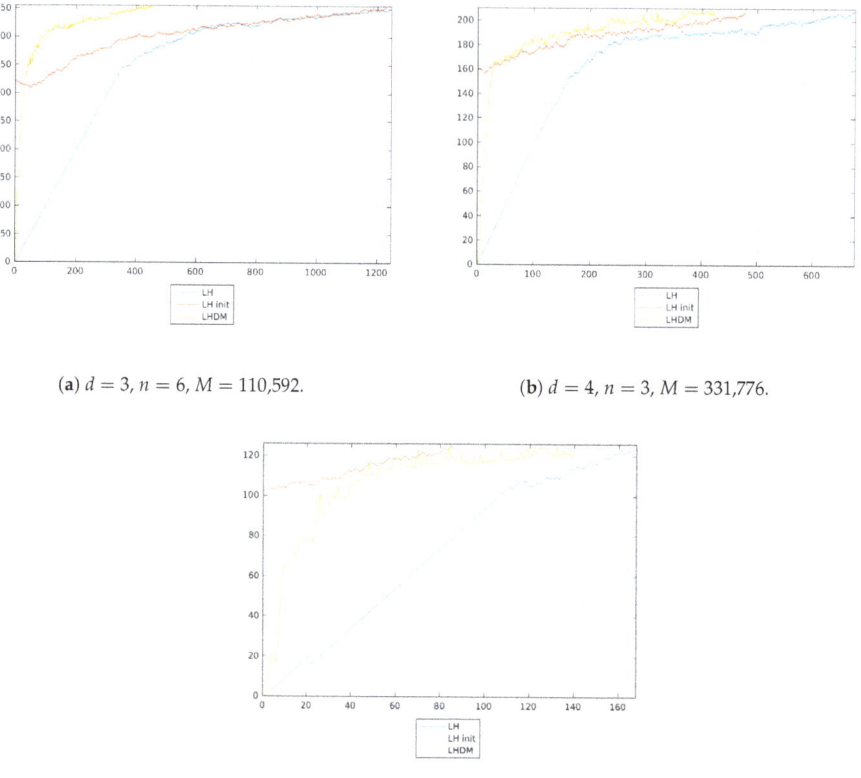

(a) $d = 3$, $n = 6$, $M = 110{,}592$.

(b) $d = 4$, $n = 3$, $M = 331{,}776$.

(c) $d = 5$, $n = 2$, $M = 1{,}048{,}576$.

Figure 2. The evolution of the cardinality of the passive set P along the iterations of the three LH algorithms for Chebyshev nodes' tests.

Table 4. Results of numerical tests on $M = (2km)^d$ Chebyshev's nodes, with $k = 4$, with different dimensions and degrees: $compr = M/mean(cpts)$ is the mean compression ratio obtained by the three methods listed; t_{LH}/t_{Titt} is the ratio between the execution time of LH and that of Titterington algorithm; t_{LH}/t_{LHDM} (t_{LHI}/t_{LHDM}) is the ratio between the execution time of LH (LHI) and that of LHDM; $cpts$ is the number of compressed Tchakaloff points and *momerr* is the final moment residual.

	Test			LH				LHI		LHDM	
d	m	M	compr	t_{LH}/t_{Titt}	t_{LH}/t_{LHDM}	cpts	momerr	t_{LHI}/t_{LHDM}	cpts	momerr	cpts	momerr
3	6	110,592	250/1	0.4/1	3.1/1	450	5.0×10^{-7}	3.5/1	450	3.4×10^{-7}	450	1.4×10^{-7}
4	3	331,776	1607/1	0.2/1	2.0/1	207	8.9×10^{-7}	3.4/1	205	9.8×10^{-7}	207	7.9×10^{-7}
5	2	1,048,576	8571/1	0.1/1	1.4/1	122	6.3×10^{-7}	1.5/1	123	3.6×10^{-7}	122	3.3×10^{-7}

4.3. Hypercubes: Low-Discrepancy Points

The direct connection of Chebyshev grids with near G-optimal designs discussed in the previous subsection suffers rapidly of the curse of dimensionality, so only regression at low degree in relatively low dimension can be treated. On the other hand, in sampling theory a number of discretization nets with good space-filling properties on hypercubes has been proposed and they allow to increase the dimension d. We refer in particular to Latin hypercube sampling or low-discrepancy points (Sobol, Halton and other popular sequences); see for example, Reference [29]. These families of points give a discrete model of hypercubes that can be used in many different deterministic and statistical applications.

Here we consider a discretization made via Halton points. We present in particular two examples, where we take as finite design space X a set of $M = 10^5$ Halton points, in $d = 4$ with regression degree $m = 5$, and in $d = 10$ with $m = 2$. In both examples, $\dim(P_{2m}^d) = \binom{2m+d}{d} = \binom{2m+d}{2m} = \binom{14}{4} = 1001$, so that the largest matrix involved in the construction is the $1001 \times 100{,}000$ Chebyshev-Vandermonde matrix C for degree $2m$ on X constructed at the beginning of the compression process (by dORTHVAND within dCATCH to compute U_{2m} in (12)).

Results are shown in Figure 3 and Table 5, using the same notation as above.

Remark 1. *The computational complexity of dCATCH mainly depends on the QR decompositions, which clearly limit the maximum size of the problem and mainly determine the execution time. Indeed, the computational complexity of a QR factorization of a matrix of size $n_r \times n_c$, with $n_c \leq n_r$, is high, namely $2(n_c^2 n_r - n_c^3/3) \approx 2n_c^2 n_r$ (see, e.g., Reference [30]).*

Titterington algorithm performs a QR factorization of a $M \times N_m$ matrix at each iteration, with the following overall computational complexity

$$C_{Titt} \approx 2\bar{k} \, M \, N_m^2,$$

where $\bar{k}$ is the number of iterations necessary for convergence, that depends on the desired G-efficiency.

On the other hand, the computational cost of one iteration of the Lawson-Hanson algorithm, fixed the passive set P, is given by the solution of an LS problem of type (15), which approximately is $2N_{2m}|P|^2$ that is the cost of a QR decomposition of a matrix of size $N_{2m} \times |P|$. However, as experimental results confirm, the evolution of the set P along the execution of the algorithm may vary significantly depending on the experiment settings, so that the exact overall complexity is hard to estimate. Lower and upper bounds are available, but may lead to heavy under- and over-estimations, respectively; cf. Reference [31] for a discussion on complexity issues.

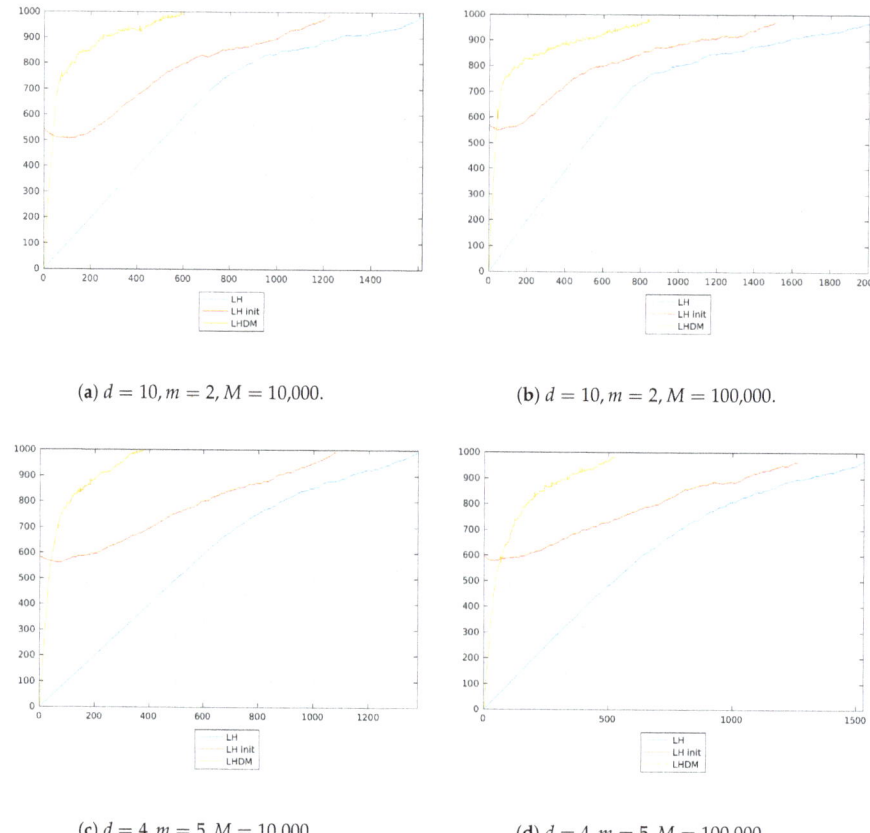

Figure 3. The evolution of the cardinality of the passive set P along the iterations of the three LH algorithms for Halton points' tests.

Table 5. Results of numerical tests on Halton points: $compr = M/mean(cpts)$ is the mean compression ratio obtained by the three methods listed; t_{LH}/t_{Titt} is the ratio between the execution time of LH and that of Titterington algorithm; t_{LH}/t_{LHDM} (t_{LHI}/t_{LHDM}) is the ratio between the execution time of LH (LHI) and that of LHDM; $cpts$ is the number of compressed Tchakaloff points and $momerr$ is the final moment residual.

Test				LH				LHI			LHDM	
d	m	M	compr	t_{LH}/t_{Titt}	t_{LH}/t_{LHDM}	cpts	momerr	t_{LHI}/t_{LHDM}	cpts	momerr	cpts	momerr
10	2	10,000	10/1	41.0/1	1.9/1	990	1.1×10^{-8}	1.9/1	988	9.8×10^{-9}	990	9.4×10^{-9}
10	2	100,000	103/1	6.0/1	3.1/1	968	3.6×10^{-7}	2.8/1	973	2.7×10^{-7}	968	4.2×10^{-7}
4	5	10,000	10/1	20.2/1	2.3/1	997	9.7×10^{-9}	2.4/1	993	1.3×10^{-8}	997	2.1×10^{-9}
4	5	100,000	103/1	2.0/1	3.8/1	969	6.6×10^{-7}	3.8/1	964	6.3×10^{-7}	969	5.3×10^{-7}

5. Conclusions

In this paper, we have presented *dCATCH* [1], a numerical software package for the computation of a d-variate near G-optimal polynomial regression design of degree m on a finite design space $X \subset \mathbb{R}^d$. The mathematical foundation is discussed connecting statistical design theoretic and approximation theoretic aspects, with a special emphasis on deterministic regression (Weighted Least Squares). The package takes advantage of an accelerated version of the classical NNLS Lawson-Hanson solver developed by the authors and applied to design compression.

As a few examples of use cases of this package we have shown the results on a complex shape (multibubble) in three dimensions, and on hypercubes discretized with Chebyshev grids and with Halton points, testing different combinations of dimensions and degrees which generate large-scale problems for a personal computer.

The present package, *dCATCH* works for any discrete measure on a discrete set X. Indeed, it could be used, other than for design compression, also in the compression of d-variate quadrature formulas, even on lower-dimensional manifolds, to give an example.

We may observe that with this approach we can compute a d-variate compressed design starting from a high-cardinality sampling set X, that discretizes a continuous compact set (see Sections 4.2 and 4.3). This design allows an m-th degree near optimal polynomial regression of a function on the whole X, by sampling on a small design support. We stress that the compressed design is function-independent and thus can be constructed "once and for all" in a pre-processing stage. This approach is potentially useful, for example, for the solution of d-variate parameter estimation problems, where we may think to model a nonlinear cost function by near optimal polynomial regression on a discrete d-variate parameter space X; cf., for example, References [32,33] for instances of parameter estimation problems from mechatronics applications (*Digital Twins* of controlled systems) and references on the subject. Minimization of the polynomial model could then be accomplished by popular methods developed in the growing research field of Polynomial Optimization, such as Lasserre's SOS (Sum of Squares) and measure-based hierarchies, and other recent methods; cf., for example, References [34–36] with the references therein.

From a computational viewpoint, the results shown in Tables 3–5 show relevant speed-ups in the compression stage, with respect to the standard Lawson-Hanson algorithm, in terms of the number of iterations required and of computing time within the Matlab scripting language. In order to further decrease the execution times and to allow us to tackle larger design problems, we would like in the near future to enrich the package *dCATCH* with an efficient C implementation of its algorithms and, possibly, a CUDA acceleration on GPUs.

Author Contributions: Investigation, M.D., F.M. and M.V. All authors have read and agreed to the published version of the manuscript.

Funding: Work partially supported by the DOR funds and the biennial project Project BIRD192932 of the University of Padova, and by the GNCS-INdAM. This research has been accomplished within the RITA "Research ITalian network on Approximation".

Conflicts of Interest: The authors declare no conflict of interest.

References

1. Dessole, M.; Marcuzzi, F.; Vianello, M. dCATCH: A Numerical Package for Compressed d-Variate Near G-Optimal Regression. Available online: https://www.math.unipd.it/~marcov/MVsoft.html (accessed on 1 June 2020).
2. Bos, L.; Vianello, M. CaTchDes: MATLAB codes for Caratheodory—Tchakaloff Near-Optimal Regression Designs. *SoftwareX* **2019**, *10*, 100349. [CrossRef]
3. Dessole, M.; Marcuzzi, F.; Vianello, M. Accelerating the Lawson-Hanson NNLS solver for large-scale Tchakaloff regression designs. *Dolomit. Res. Notes Approx. DRNA* **2020**, *13*, 20–29.
4. Atkinson, A.; Donev, A.; Tobias, R. *Optimum Experimental Designs, with SAS*; Oxford University Press: Oxford, UK, 2007.
5. Pukelsheim, F. *Optimal Design of Experiments*; SIAM: Philadelphia, PA, USA, 2006.
6. Celant, G.; Broniatowski, M. *Interpolation and Extrapolation Optimal Designs 2-Finite Dimensional General Models*; Wiley: Hoboken, NJ, USA, 2017.
7. Mandal, A.; Wong, W.K.; Yu, Y. Algorithmic searches for optimal designs. In *Handbook of Design and Analysis of Experiments*; CRC Press: Boca Raton, FL, USA, 2015; pp. 755–783.

8. De Castro, Y.; Gamboa, F.; Henrion, D.; Hess, R.; Lasserre, J.B. Approximate optimal designs for multivariate polynomial regression. *Ann. Stat.* **2019**, *47*, 127–155. [CrossRef]
9. Dette, H.; Pepelyshev, A.; Zhigljavsky, A. Improving updating rules in multiplicative algorithms for computing D-optimal designs. *Comput. Stat. Data Anal.* **2008**, *53*, 312–320. [CrossRef]
10. Torsney, B.; Martin-Martin, R. Multiplicative algorithms for computing optimum designs. *J. Stat. Plan. Infer.* **2009**, *139*, 3947–3961. [CrossRef]
11. Bloom, T.; Bos, L.; Levenberg, N.; Waldron, S. On the Convergence of Optimal Measures. *Constr. Approx.* **2008**, *32*, 159–169. [CrossRef]
12. Bos, L. Some remarks on the Fejér problem for lagrange interpolation in several variables. *J. Approx. Theory* **1990**, *60*, 133–140. [CrossRef]
13. Calvi, J.P.; Levenberg, N. Uniform approximation by discrete least squares polynomials. *J. Approx. Theory* **2008**, *152*, 82–100. [CrossRef]
14. Kiefer, J.; Wolfowitz, J. The equivalence of two extremum problems. *Can. J. Math.* **1960**, *12*, 363–366. [CrossRef]
15. Putinar, M. A note on Tchakaloff's theorem. *Proc. Am. Math. Soc.* **1997**, *125*, 2409–2414. [CrossRef]
16. Titterington, D. Algorithms for computing D-optimal designs on a finite design space. In *Proceedings of the 1976 Conference on Information Science and Systems*; John Hopkins University: Baltimore, MD, USA, 1976; Volume 3, pp. 213–216.
17. Burkardt, J. MONOMIAL: A Matlab Library for Multivariate Monomials. Available online: https://people.sc.fsu.edu/~jburkardt/m_src/monomial/monomial.html (accessed on 1 June 2020).
18. Bos, L.; Piazzon, F.; Vianello, M. Near optimal polynomial regression on norming meshes. In *Sampling Theory and Applications 2019*; IEEE Xplore Digital Library: New York, NY, USA, 2019.
19. Bos, L.; Piazzon, F.; Vianello, M. Near G-optimal Tchakaloff designs. *Comput. Stat.* **2020**, *35*, 803–819. [CrossRef]
20. Tchakaloff, V. Formules de cubatures mécaniques à coefficients non négatifs. *Bull. Sci. Math.* **1957**, *81*, 123–134.
21. Carathéodory, C. Über den Variabilitätsbereich der Fourier'schen Konstanten von positiven harmonischen Funktionen. *Rendiconti Del Circolo Matematico di Palermo (1884–1940)* **1911**, *32*, 193–217. [CrossRef]
22. Sommariva, A.; Vianello, M. Compression of Multivariate Discrete Measures and Applications. *Numer. Funct. Anal. Optim.* **2015**, *36*, 1198–1223. [CrossRef]
23. Pleśniak, W. Multivariate Jackson Inequality. *J. Comput. Appl. Math.* **2009**, *233*, 815–820. [CrossRef]
24. Lawson, C.L.; Hanson, R.J. *Solving Least Squares Problems*; SIAM: Philadelphia, PA, USA, 1995; Volume 15.
25. Van Benthem, M.H.; Keenan, M.R. Fast algorithm for the solution of large-scale non-negativity-constrained least squares problems. *J. Chemom.* **2004**, *18*, 441–450. [CrossRef]
26. Hardin, D.; Michaels, T.; Saff, E. A Comparison of Popular Point Configurations on S^2. *Dolomit. Res. Notes Approx. DRNA* **2016**, *9*, 16–49.
27. Bloom, T.; Bos, L.; Calvi, J.; Levenberg, N. Polynomial Interpolation and Approximation in $\mathbb{C}^d$. *Ann. Polon. Math.* **2012**, *106*, 53–81. [CrossRef]
28. De Marchi, S.; Piazzon, F.; Sommariva, A.; Vianello, M. Polynomial Meshes: Computation and Approximation. In Proceedings of the CMMSE 2015, Rota Cadiz, Spain, 6–10 July 2015; pp. 414–425.
29. Dick, J.; Pillichshammer, F. *Digital Nets and Sequences-Discrepancy Theory and Quasi—Monte Carlo Integration*; Cambridge University Press: Cambridge, UK, 2010.
30. Golub, G.H.; Van Loan, C.F. *Matrix Computations*, 3rd ed.; Johns Hopkins University Press: Baltimore, MD, USA, 1996.
31. Slawski, M. Nonnegative Least Squares: Comparison of Algorithms. Available online: https://sites.google.com/site/slawskimartin/code (accessed on 1 June 2020).
32. Beghi, A.; Marcuzzi, F.; Martin, P.; Tinazzi, F.; Zigliotto, M. Virtual prototyping of embedded control software in mechatronic systems: A case study. *Mechatronics* **2017**, *43*, 99–111. [CrossRef]
33. Beghi, A.; Marcuzzi, F.; Rampazzo, M. A Virtual Laboratory for the Prototyping of Cyber-Physical Systems. *IFAC-PapersOnLine* **2016**, *49*, 63–68. [CrossRef]
34. Lasserre, J.B. The moment-SOS hierarchy. *Proc. Int. Cong. Math.* **2018**, *4*, 3791–3814.

35. De Klerk, E.; Laurent, M. A survey of semidefinite programming approaches to the generalized problem of moments and their error analysis. In *World Women in Mathematics 2018-Association for Women in Mathematics Series*; Springer: Cham, Switzerland, 2019; Volume 20, pp. 17–56.
36. Martinez, A.; Piazzon, F.; Sommariva, A.; Vianello, M. Quadrature-based polynomial optimization. *Optim. Lett.* **2020**, *35*, 803–819. [CrossRef]

© 2020 by the authors. Licensee MDPI, Basel, Switzerland. This article is an open access article distributed under the terms and conditions of the Creative Commons Attribution (CC BY) license (http://creativecommons.org/licenses/by/4.0/).

Article

Exact Solutions to the Maxmin Problem max $\|Ax\|$ Subject to $\|Bx\| \leq 1$

Soledad Moreno-Pulido [1,†], Francisco Javier Garcia-Pacheco [1,†], Clemente Cobos-Sanchez [2,†] and Alberto Sanchez-Alzola [3,*,†]

1. Department of Mathematics, College of Engineering, University of Cadiz, 11510 Puerto Real, Spain; soledad.moreno@uca.es (S.M.-P.); garcia.pacheco@uca.es (F.J.G.-P.)
2. Department of Electronics, College of Engineering, University of Cadiz, 11510 Puerto Real, Spain; clemente.cobos@uca.es
3. Department of Statistics and Operation Research, College of Engineering, University of Cadiz, 11510 Puerto Real, Spain
* Correspondence: alberto.sanchez@gm.uca.es
† These authors contributed equally to this work.

Received: 15 October 2019; Accepted: 30 December 2019; Published: 4 January 2020

Abstract: In this manuscript we provide an exact solution to the maxmin problem max $\|Ax\|$ subject to $\|Bx\| \leq 1$, where A and B are real matrices. This problem comes from a remodeling of max $\|Ax\|$ subject to min $\|Bx\|$, because the latter problem has no solution. Our mathematical method comes from the Abstract Operator Theory, whose strong machinery allows us to reduce the first problem to max $\|Cx\|$ subject to $\|x\| \leq 1$, which can be solved exactly by relying on supporting vectors. Finally, as appendices, we provide two applications of our solution: first, we construct a truly optimal minimum stored-energy Transcranial Magnetic Stimulation (TMS) coil, and second, we find an optimal geolocation involving statistical variables.

Keywords: maxmin; supporting vector; matrix norm; TMS coil; optimal geolocation

MSC: 47L05, 47L90, 49J30, 90B50

---

1. Introduction

1.1. Scope

Different scientific fields, such as Physics, Statistics, Economics, or Engineering, deal with real-life problems that are usually modelled by the experts on those fields using matrices and their norms (see [1–6]). A typical modelling is the following original maxmin problem

$$\begin{cases} \max \|Ax\| \\ \min \|Bx\|. \end{cases}$$

One of the most iconic results in this manuscript (Theorem 2) shows that the previous problem, regarded strictly as a multiple optimization problem, has no solutions. To save this obstacle we provide a different model, such as

$$\begin{cases} \max \|Ax\| \\ \|Bx\| \leq 1. \end{cases}$$

Here in this article we justify the remodelling of the original maxmin problem and we solve it by making use of supporting vectors. This concept comes from the Theory of Banach Spaces and Operator

Theory. Given a matrix A, a supporting vector is a unit vector x such that A attains its norm at x, that is, x is a solution of the following single optimization problem:

$$\begin{cases} \max \|Ax\| \\ \|x\| = 1. \end{cases}$$

The geometric and topological structure of supporting vectors can be consulted in [7–9]. On the other hand, generalized supporting vectors are defined and studied in [7,8]. The generalized supporting vectors of a finite sequence of matrices $A_1, \ldots, A_n$, for the Euclidean norm $\|\bullet\|_2$, are the solutions of

$$\begin{cases} \max \|A_1 x\|_2^2 + \cdots + \|A_n x\|_2^2 \\ \|x\|_2 = 1. \end{cases}$$

This optimization problem clearly generalizes the previous one.

Supporting vectors were originally applied in [10] to truly optimally design a TMS coil, because until that moment TMS coils had only been designed by means of heuristic methods, which were never proved to be convergent. In [10] a three-component TMS coil problem is posed but only the one-component case was resolved. The three-component case was stated and solved by means of the generalized supporting vectors in [8]. In this manuscript, we model a TMS coil with a maxmin problem and solve it exactly with our method.

A second application of supporting vectors was given in [8], where an optimal location situation using Principal Component Analysis (PCA) was solved. In this manuscript, we model a more complex PCA problem as an optimal maxmin geolocation involving statistical variables.

For other perspective on supporting vectors and generalized supporting vectors, we refer the reader to [9].

1.2. Background

In the first place, we refer the reader to [8] (Preliminaries) for a general review of multiobjective optimization problems and their reformulations to avoid the lack of solutions (generally caused by the existence of many objective functions).

The original maxmin optimization problem has the form

$$M := \begin{cases} \max g(x) \\ \min f(x) \end{cases}$$

where $f, g : X \to (0, \infty)$ are real-valued functions and X is a nonempty set. Notice that

$$\text{sol}(M) = \arg\max g(x) \cap \arg\min f(x).$$

Many real-life problems can be mathematically model, such as a maxmin. However, this kind of multiobjective optimization problems may have the inconvenience of lacking a solution. If this occurs, then we are in need of remodeling the real-life problem with another mathematical optimization problem that has a solution and still models the real-life problem very accurately.

According to [10] (Theorem 5.1), one can realize that, in case $\text{sol}(M) = \emptyset$, the following optimization problems are good alternatives to keep modeling the real-life problem accurately:

- $\begin{cases} \max g(x) \\ \min f(x) \end{cases} \xrightarrow{\text{reform}} \begin{cases} \min \frac{f(x)}{g(x)} \\ g(x) \neq 0 \end{cases}$.

- $\begin{cases} \max g(x) \\ \min f(x) \end{cases} \xrightarrow{\text{reform}} \begin{cases} \max \frac{g(x)}{f(x)} \\ f(x) \neq 0 \end{cases}$.

- $\begin{cases} \max g(x) \\ \min f(x) \end{cases} \xrightarrow{\text{reform}} \begin{cases} \max g(x) \\ f(x) \leq a \end{cases}$.

- $\begin{cases} \max g(x) \\ \min f(x) \end{cases} \xrightarrow{\text{reform}} \begin{cases} \min f(x) \\ g(x) \geq b \end{cases}$.

We will prove in the third section that all four previous reformulations are equivalent for the original maxmin $\begin{cases} \max \|Ax\| \\ \min \|Bx\| \end{cases}$. In the fourth section, we will solve the reformulation $\begin{cases} \max \|Ax\| \\ \|Bx\| \leq 1 \end{cases}$.

2. Characterizations of Operators with Null Kernel

Kernels will play a fundamental role towards solving the general reformulated maxmin (2) as shown in the next section. This is why we first study the operators with null kernel.

Throughout this section, all monoid actions considered will be left, all rings will be associative, all rings will be unitary rngs, all absolute semi-values and all semi-norms will be non-zero, all modules over rings will be unital, all normed spaces will be real or complex and all algebras will be unitary and complex.

Given a rng R and an element $s \in R$, we will denote by $\ell d(s)$ to the set of left divisors of s, that is,

$$\ell d(s) := \{r \in R : \exists t \in R \setminus \{0\} \text{ with } rt = s\}.$$

Similarly, $rd(s)$ stands for the set of right divisors of s. If R is a ring, then the set of its invertibles is usually denoted by $\mathcal{U}(R)$. Notice that $\ell d(1)$ ($rd(1)$) is precisely the subset of elements of R which are right-(left) invertible. As a consequence, $\mathcal{U}(R) = \ell d(1) \cap rd(1)$. Observe also that $\ell d(0) \cap rd(1) = \emptyset = rd(0) \cap \ell d(1)$. In general we have that $\ell d(0) \cap \ell d(1) \neq \emptyset$ and $rd(0) \cap rd(1) \neq \emptyset$. Later on in Example 1 we will provide an example of a ring where $rd(0) \cap rd(1) \neq \emptyset$.

Recall that an element p of a monoid is called involutive if $p^2 = 1$. Given a rng R, an involution is an additive, antimultiplicative, composition-involutive map $* : R \to R$. A $*$-rng is a rng endowed with an involution.

The categorical concept of monomorphism will play an important role in this manuscript. A morphism $f \in \hom_{\mathcal{C}}(A, B)$ between objects A and B in a category $\mathcal{C}$ is said to be a monomorphism provided that $f \circ g = f \circ h$ implies $g = h$ for all $C \in \text{ob}(\mathcal{C})$ and all $g, h \in \hom_{\mathcal{C}}(C, A)$. Once can check that if $f \in \hom_{\mathcal{C}}(A, B)$ and there exist $C_0 \in \text{ob}(\mathcal{C})$ and $g_0 \in \hom_{\mathcal{C}}(B, C_0)$ such that $g_0 \circ f$ is a monomorphism, then f is also a monomorphism. In particular, if $f \in \hom_{\mathcal{C}}(A, B)$ is a section, that is, exists $g \in \hom_{\mathcal{C}}(B, A)$ such that $g \circ f = I_A$, then f is a monomorphism. As a consequence, the elements of $\hom_{\mathcal{C}}(A, A)$ that have a left inverse are monomorphisms. In some categories, the last condition suffices to characterize monomorphisms. This is the case, for instance, of the category of vector spaces over a division ring.

Recall that $\mathcal{CL}(X, Y)$ denotes the space of continuous linear operators from a topological vector space X to another topological vector space Y.

Proposition 1. *A continuous linear operator $T : X \to Y$ between locally convex Hausdorff topological vector spaces X and Y verifies that $\ker(T) \neq \{0\}$ if and only if exists $S \in \mathcal{CL}(Y, X) \setminus \{0\}$ with $T \circ S = 0$. In particular, if $X = Y$, then $\ker(T) \neq \{0\}$ if and only if $T \in \ell d(0)$ in $\mathcal{CL}(X)$.*

Proof. Let $S \in \mathcal{CL}(Y, X) \setminus \{0\}$ such that $T \circ S = 0$. Fix any $y \in Y \setminus \ker(S)$, then $S(y) \neq 0$ and $T(S(y)) = 0$ so $S(y) \in \ker(T) \setminus \{0\}$. Conversely, if $\ker(T) \neq \{0\}$, then fix $x_0 \in \ker(T) \setminus \{0\}$ and $y_0^* \in Y^* \setminus \{0\}$ (the existence of y^* is guaranteed by the Hahn-Banach Theorem on the Hausdorff locally convex topological vector space Y). Next, consider

$$S : Y \to X$$
$$y \mapsto S(y) := y_0^*(y) x_0.$$

Notice that $S \in \mathcal{CL}(Y, X) \setminus \{0\}$ and $T \circ S = 0$. □

Theorem 1. *Let $T : X \to Y$ be a continuous linear operator between locally convex Hausdorff topological vector spaces X and Y. Then:*

1. *If T is a section, then $\ker(T) = \{0\}$*
2. *In case X and Y are Banach spaces, $T(X)$ is topologically complemented in Y and $\ker(T) = \{0\}$, then T is a section.*

Proof.

1. Trivial since sections are monomorphisms.
2. Consider $T : X \to T(X)$. Since $T(X)$ is topologically complemented in Y we have that $T(X)$ is closed in Y, thus it is a Banach space. Therefore, the Open Mapping Theorem assures that $T : X \to T(X)$ is an isomorphism. Let $T^{-1} : T(X) \to X$ be the inverse of $T : X \to T(X)$. Now consider $P : Y \to Y$ to be a continuous linear projection such that $P(Y) = T(X)$. Finally, it suffices to define $S := T^{-1} \circ P$ since $S \circ T = I_X$.

□

We will finalize this section with a trivial example of a matrix $A \in \mathbb{R}^{3\times 2}$ such that $A \in rd(I) \cap rd(0)$.

Example 1. *Consider*

$$A = \begin{pmatrix} 1 & 0 \\ 0 & 1 \\ 0 & 0 \end{pmatrix}.$$

It is not hard to check that $\ker(A) = \{(0,0)\}$ thus A is left-invertible by Theorem 1(2) and so $A \in rd(I)$. In fact,

$$\begin{pmatrix} 1 & 0 & 0 \\ 0 & 1 & 0 \end{pmatrix} \begin{pmatrix} 1 & 0 \\ 0 & 1 \\ 0 & 0 \end{pmatrix} = \begin{pmatrix} 1 & 0 \\ 0 & 1 \end{pmatrix}.$$

Finally,

$$\begin{pmatrix} 0 & 0 & 1 \\ 0 & 0 & 1 \end{pmatrix} \begin{pmatrix} 1 & 0 \\ 0 & 1 \\ 0 & 0 \end{pmatrix} = \begin{pmatrix} 0 & 0 \\ 0 & 0 \end{pmatrix}.$$

3. Remodeling the Original Maxmin Problem max $\|T(x)\|$ Subject to min $\|S(x)\|$

3.1. The Original Maxmin Problem Has No Solutions

This subsection begins with the following theorem:

Theorem 2. *Let $T, S : X \to Y$ be nonzero continuous linear operators between Banach spaces X and Y. Then the original maxmin problem*

$$\begin{cases} \max \|T(x)\| \\ \min \|S(x)\| \end{cases} \tag{1}$$

has trivially no solution.

Proof. Observe that $\arg\min \|S(x)\| = \ker(S)$ and $\arg\max \|T(x)\| = \emptyset$ because $T \neq \{0\}$. Then the set of solutions of Problem (1) is

$$\arg\min \|S(x)\| \cap \arg\max \|T(x)\| = \ker(S) \cap \emptyset = \emptyset.$$

□

As a consequence, Problem (1) must be reformulated or remodeled.

3.2. Equivalent Reformulations for the Original Maxmin Problem

According to the Background section, we begin with the following reformulation:

$$\begin{cases} \max \|T(x)\| \\ \|S(x)\| \leq 1 \end{cases} \quad (2)$$

Please note that $\arg\max_{\|S(x)\|\leq 1} \|T(x)\|$ is a $\mathbb{K}$-symmetric set, where $\mathbb{K} := \mathbb{R}$ or $\mathbb{C}$, in other words, if $\lambda \in \mathbb{K}$ and $|\lambda| = 1$, then $\lambda x \in \arg\max_{\|S(x)\|\leq 1} \|T(x)\|$ for every $x \in \arg\max_{\|S(x)\|\leq 1} \|T(x)\|$. The finite dimensional version of the previous reformulation is

$$\begin{cases} \max \|Ax\| \\ \|Bx\| \leq 1 \end{cases} \quad (3)$$

where $A, B \in \mathbb{R}^{m \times n}$.

Recall that $\mathcal{B}(X, Y)$ denotes the space of bounded operators from X to Y.

Lemma 1. *Let $T, S \in \mathcal{B}(X, Y)$ where X and Y are Banach spaces. If the general reformulated maxmin problem*

$$\begin{cases} \max \|T(x)\| \\ \|S(x)\| \leq 1 \end{cases}$$

has a solution, then $\ker(S) \subseteq \ker(T)$.

Proof. If $\ker(S) \setminus \ker(T) \neq \emptyset$, then it suffices to consider the sequence $(nx_0)_{n\in\mathbb{N}}$ for $x_0 \in \ker(S) \setminus \ker(T)$, since $\|S(nx_0)\| = 0 \leq 1$ for all $n \in \mathbb{N}$ and $\|T(nx_0)\| = n\|T(x_0)\| \to \infty$ as $n \to \infty$. □

The general maxmin (1) can also be reformulated as

$$\begin{cases} \max \|T(x)\| \\ \min \|S(x)\| \end{cases} \xrightarrow{\text{reform}} \begin{cases} \max \frac{\|T(x)\|}{\|S(x)\|} \\ \|S(x)\| \neq 0 \end{cases}$$

Lemma 2. *Let $T, S \in \mathcal{B}(X, Y)$ where X and Y are Banach spaces. If the second general reformulated maxmin problem*

$$\begin{cases} \max \frac{\|T(x)\|}{\|S(x)\|} \\ \|S(x)\| \neq 0 \end{cases}$$

has a solution, then $\ker(S) \subseteq \ker(T)$.

Proof. Suppose there exists $x_0 \in \ker(S) \setminus \ker(T)$. Then fix an arbitrary $x_1 \in X \setminus \ker(S)$. Notice that

$$\frac{\|T(nx_0 + x_1)\|}{\|S(nx_0 + x_1)\|} \geq \frac{n\|T(x_0)\| - \|T(x_1)\|}{\|S(x_1)\|} \to \infty$$

as $n \to \infty$. □

The next theorem shows that the previous two reformulations are in fact equivalent.

Theorem 3. *Let $T, S \in \mathcal{B}(X,Y)$ where X and Y are Banach spaces. Then*

$$\bigcup_{t>0} t \arg\max_{\|S(x)\|\leq 1} \|T(x)\| = \arg\max_{\|S(x)\|\neq 0} \frac{\|T(x)\|}{\|S(x)\|}.$$

Proof. Let $x_0 \in \arg\max_{\|S(x)\|\leq 1} \|T(x)\|$ and $t_0 > 0$. Fix an arbitrary $y \in X \setminus \ker(S)$. Notice that $x_0 \notin \ker(S)$ in virtue of Theorem 1. Then

$$\|T(x_0)\| \geq \left\|T\left(\frac{y}{\|S(y)\|}\right)\right\|,$$

therefore

$$\frac{\|T(tx_0)\|}{\|S(tx_0)\|} = \frac{\|T(x_0)\|}{\|S(x_0)\|} \geq \|T(x_0)\| \geq \left\|T\left(\frac{y}{\|S(y)\|}\right)\right\|.$$

Conversely, let $x_0 \in \arg\max_{\|S(x)\|\neq 0} \frac{\|T(x)\|}{\|S(x)\|}$. Fix an arbitrary $y \in X$ with $\|S(y)\| \leq 1$. Then

$$\left\|T\left(\frac{x_0}{\|S(x_0)\|}\right)\right\| = \frac{\|T(x_0)\|}{\|S(x_0)\|} \geq \frac{\|T(y)\|}{\|S(y)\|} \geq \|T(y)\|$$

which means that

$$\frac{x_0}{\|S(x_0)\|} \in \arg\max_{\|S(x)\|\leq 1} \|T(x)\|$$

and thus

$$x_0 \in \|S(x_0)\| \arg\max_{\|S(x)\|\leq 1} \|T(x)\| \subseteq \bigcup_{t>0} t \arg\max_{\|S(x)\|\leq 1} \|T(x)\|.$$

□

The reformulation

$$\begin{cases} \min \frac{\|S(x)\|}{\|T(x)\|} \\ \|T(x)\| \neq 0 \end{cases}$$

is slightly different from the previous two reformulations. In fact, if $\ker(S) \setminus \ker(T) \neq \varnothing$, then $\arg\min_{\|T(x)\|\neq 0} \frac{\|S(x)\|}{\|T(x)\|} = \ker(S) \setminus \ker(T)$. The previous reformulation is equivalent to the following one as shown in the next theorem:

$$\begin{cases} \min \|S(x)\| \\ \|T(x)\| \geq 1 \end{cases}$$

Theorem 4. *Let $T, S \in \mathcal{B}(X,Y)$ where X and Y are Banach spaces. Then*

$$\bigcup_{t>0} t \arg\min_{\|T(x)\|\geq 1} \|S(x)\| = \arg\min_{\|T(x)\|\neq 0} \frac{\|S(x)\|}{\|T(x)\|}.$$

We spare of the details of the proof of the previous theorem to the reader. Notice that if $\ker(S) \setminus \ker(T) \neq \varnothing$, then $\arg\min_{\|T(x)\|\geq 1} \|S(x)\| = \ker(S) \setminus \{x \in\colon \|T(x)\| < 1\}$. However, if $\ker(S) \subseteq \ker(T)$, then all four reformulations are equivalent, as shown in the next theorem, whose proof's details we spare again to the reader.

Theorem 5. *Let $T, S \in \mathcal{B}(X,Y)$ where X and Y are Banach spaces. If $\ker(S) \subseteq \ker(T)$, then*

$$\arg\max_{\|S(x)\|\neq 0} \frac{\|T(x)\|}{\|S(x)\|} = \arg\min_{\|T(x)\|\neq 0} \frac{\|S(x)\|}{\|T(x)\|}.$$

4. Solving the Maxmin Problem max $\|T(x)\|$ Subject to $\|S(x)\| \leq 1$

We will distinguish between two cases.

4.1. First Case: S Is an Isomorphism Over Its Image

By bearing in mind Theorem 5, we can focus on the first reformulation proposed at the beginning of the previous section:

$$\begin{cases} \max \|T(x)\| \\ \min \|S(x)\| \end{cases} \xrightarrow{\text{reform}} \begin{cases} \max \|T(x)\| \\ \|S(x)\| \leq 1 \end{cases}$$

The idea we propose to solve the previous reformulation is to make use of supporting vectors (see [7–10]). Recall that if $R : X \to Y$ is a continuous linear operator between Banach spaces, then the set of supporting vectors of R is defined by

$$\operatorname{suppv}(R) := \arg\max_{\|x\| \leq 1} \|R(x)\|.$$

The idea of using supporting vectors is that the optimization problem

$$\begin{cases} \max \|R(x)\| \\ \|x\| \leq 1 \end{cases}$$

whose solutions are by definition the supporting vectors of R, can be easily solved theoretically and computationally (see [8]).

Our first result towards this direction considers the case where S is an isomorphism over its image.

Theorem 6. *Let $T, S \in \mathcal{B}(X,Y)$ where X and Y are Banach spaces. Suppose that S is an isomorphism over its image and $S^{-1} : S(X) \to X$ denotes its inverse. Suppose also that $S(X)$ is complemented in Y, being $p : Y \to Y$ a continuous linear projection onto $S(X)$. Then*

$$S^{-1}\left(S(X) \cap \arg\max_{\|y\| \leq 1}\left\|\left(T \circ S^{-1} \circ p\right)(y)\right\|\right) \subseteq \arg\max_{\|S(x)\| \leq 1} \|T(x)\|.$$

If, in addition, $\|p\| = 1$, then

$$\arg\max_{\|S(x)\| \leq 1} \|T(x)\| = S^{-1}\left(S(X) \cap \arg\max_{\|y\| \leq 1}\left\|\left(T \circ S^{-1} \circ p\right)(y)\right\|\right).$$

Proof. We will show first that

$$S(X) \cap \arg\max_{\|y\| \leq 1}\left\|\left(T \circ S^{-1} \circ p\right)(y)\right\| \subseteq S\left(\arg\max_{\|S(x)\| \leq 1} \|T(x)\|\right).$$

Let $y_0 = S(x_0) \in \arg\max_{\|y\| \leq 1}\left\|\left(T \circ S^{-1} \circ p\right)(y)\right\|$. We will show that $x_0 \in \arg\max_{\|S(x)\| \leq 1} \|T(x)\|$. Indeed, let $x \in X$ with $\|S(x)\| \leq 1$. Since $\|S(x_0)\| = \|y_0\| \leq 1$, by assumption we obtain

$$\begin{aligned} \|T(x)\| &= \left\|\left(T \circ S^{-1} \circ p\right)(S(x))\right\| \\ &\leq \left\|\left(T \circ S^{-1} \circ p\right)(y_0)\right\| \\ &= \left\|\left(T \circ S^{-1} \circ p\right)(S(x_0))\right\| \\ &= \|T(x_0)\|. \end{aligned}$$

Now assume that $\|p\|=1$. We will show that

$$S\left(\arg\max_{\|S(x)\|\leq 1}\|T(x)\|\right) \subseteq S(X) \cap \arg\max_{\|y\|\leq 1}\left\|\left(T\circ S^{-1}\circ p\right)(y)\right\|.$$

Let $x_0 \in \arg\max_{\|S(x)\|\leq 1}\|T(x)\|$, we will show that $S(x_0) \in \arg\max_{\|y\|\leq 1}\left\|\left(T\circ S^{-1}\circ p\right)(y)\right\|$. Indeed, let $y \in B_Y$. Observe that

$$\left\|S\left(S^{-1}(p(y))\right)\right\| = \|p(y)\| \leq \|y\| \leq 1$$

so by assumption

$$\begin{aligned}
\left\|\left(T\circ S^{-1}\circ p\right)(y)\right\| &= \left\|T\left(S^{-1}(p(y))\right)\right\| \\
&\leq \|T(x_0)\| \\
&= \left\|T\left(S^{-1}(p(S(x_0)))\right)\right\| \\
&= \left\|\left(T\circ S^{-1}\circ p\right)(S(x_0))\right\|.
\end{aligned}$$

□

Notice that, in the settings of Theorem 6, $S^{-1}\circ p$ is a left-inverse of S, in other words, S is a section, as in Theorem 1(2).

Taking into consideration that every closed subspace of a Hilbert space is 1-complemented (see [11,12] to realize that this fact characterizes Hilbert spaces of dimension ≥ 3), we directly obtain the following corollary.

Corollary 1. *Let $T, S \in \mathcal{B}(X, Y)$ where X is a Banach space and Y a Hilbert space. Suppose that S is an isomorphism over its image and let $S^{-1} : S(X) \to X$ be its inverse. Then*

$$\begin{aligned}
\arg\max_{\|S(x)\|\leq 1}\|T(x)\| &= S^{-1}\left(S(X) \cap \arg\max_{\|y\|\leq 1}\left\|\left(T\circ S^{-1}\circ p\right)(y)\right\|\right) \\
&= S^{-1}\left(S(X) \cap \operatorname{suppv}\left(T\circ S^{-1}\circ p\right)\right)
\end{aligned}$$

where $p : Y \to Y$ is the orthogonal projection on $S(X)$.

4.2. The Moore–Penrose Inverse

If $B \in \mathbb{K}^{m\times n}$, then the Moore–Penrose inverse of B, denoted by B^+, is the only matrix $B^+ \in \mathbb{K}^{n\times m}$ which verifies the following:

- $B = BB^+B$.
- $B^+ = B^+BB^+$.
- $BB^+ = (BB^+)^*$.
- $B^+B = (B^+B)^*$.

If $\ker(B) = 0$, then B^+ is a left-inverse of B. Even more, BB^+ is the orthogonal projection onto the range of B, thus we have the following result from Corollary 1.

Corollary 2. *Let $A, B \in \mathbb{R}^{m\times n}$ such that $\ker(B) = \{0\}$. Then*

$$\begin{aligned}
B\left(\arg\max_{\|Bx\|_2\leq 1}\|Ax\|_2\right) &= B\mathbb{R}^n \cap \arg\max_{\|y\|_2\leq 1}\|AB^+y\|_2 \\
&= B\mathbb{R}^n \cap \operatorname{suppv}\left(AB^+\right)
\end{aligned}$$

According to the previous Corollary, in its settings, if $y_0 \in \arg\max_{\|y\|_2 \leq 1} \|AB^+ y\|_2$ and there exists $x_0 \in \mathbb{R}^n$ such that $y_0 = Bx_0$, then $x_0 \in \arg\max_{\|Bx\|_2 \leq 1} \|Ax\|_2$ and x_0 can be computed as

$$x_0 = B^+ B x_0 = B^+ y_0.$$

4.3. Second Case: S Is Not an Isomorphism Over Its Image

What happens if S is not an isomorphism over its image? Next theorem answers this question.

Theorem 7. *Let $T, S \in \mathcal{B}(X, Y)$ where X and Y are Banach spaces. Suppose that $\ker(S) \subseteq \ker(T)$. If*

$$\pi : X \to X/\ker(S)$$
$$x \mapsto \pi(x) := x + \ker(S)$$

denotes the quotient map, then

$$\arg\max_{\|S(x)\| \leq 1} \|T(x)\| = \pi^{-1}\left(\arg\max_{\|\overline{S}(\pi(x))\| \leq 1} \|\overline{T}(\pi(x))\|\right),$$

where

$$\overline{T} : \frac{X}{\ker(S)} \to Y$$
$$\pi(x) \mapsto \overline{T}(\pi(x)) := T(x)$$

and

$$\overline{S} : \frac{X}{\ker(S)} \to Y$$
$$\pi(x) \mapsto \overline{S}(\pi(x)) := S(x).$$

Proof. Let $x_0 \in \arg\max_{\|S(x)\| \leq 1} \|T(x)\|$. Fix an arbitrary $y \in X$ with $\|\overline{S}(\pi(y))\| \leq 1$. Then $\|S(y)\| = \|\overline{S}(\pi(y))\| \leq 1$ therefore

$$\|\overline{T}(\pi(x_0))\| = \|T(x_0)\| \geq \|T(y)\| = \|\overline{T}(\pi(y))\|.$$

This shows that $\pi(x_0) \in \arg\max_{\|\overline{S}(\pi(x))\| \leq 1} \|\overline{T}(\pi(x))\|$. Conversely, let

$$\pi(x_0) \in \arg\max_{\|\overline{S}(\pi(x))\| \leq 1} \|\overline{T}(\pi(x))\|.$$

Fix an arbitrary $y \in X$ with $\|S(y)\| \leq 1$. Then $\|\overline{S}(\pi(y))\| = \|S(y)\| \leq 1$ therefore

$$\|T(x_0)\| = \|\overline{T}(\pi(x_0))\| \geq \|\overline{T}(\pi(y))\| = \|T(y)\|.$$

This shows that $x_0 \in \arg\max_{\|S(x)\| \leq 1} \|T(x)\|$. □

Please note that in the settings of Theorem 7, if $S(X)$ is closed in Y, then $\overline{S}$ is an isomorphism over its image $S(X)$, and thus in this case Theorem 7 reduces the reformulated maxmin to Theorem 6.

4.4. Characterizing When the Finite Dimensional Reformulated Maxmin Has a Solution

The final part of this section is aimed at characterizing when the finite dimensional reformulated maxmin has a solution.

Lemma 3. *Let $S : X \to Y$ be a bounded operator between finite dimensional Banach spaces X and Y. If $(x_n)_{n \in \mathbb{N}}$ is a sequence in $\{x \in X : \|S(x)\| \leq 1\}$, then there is a sequence $(z_n)_{n \in \mathbb{N}}$ in $\ker(S)$ so that $(x_n + z_n)_{n \in \mathbb{N}}$ is bounded.*

Proof. Consider the linear operator

$$\overline{S}: \quad \frac{X}{\ker(S)} \to Y$$
$$x + \ker(S) \mapsto \overline{S}(x + \ker(S)) = S(x).$$

Please note that

$$\|\overline{S}(x_n + \ker(S))\| = \|S(x_n)\| \le 1$$

for all $n \in \mathbb{N}$, therefore the sequence $(x_n + \ker(S))_{n \in \mathbb{N}}$ is bounded in $\frac{X}{\ker(S)}$ because $\frac{X}{\ker(S)}$ is finite dimensional and $\overline{S}$ has null kernel so its inverse is continuous. Finally, choose $z_n \in \ker(S)$ such that $\|x_n + z_n\| < \|x_n + \ker(S)\| + \frac{1}{n}$ for all $n \in \mathbb{N}$. □

Lemma 4. *Let $A, B \in \mathbb{R}^{m \times n}$. If $\ker(B) \subseteq \ker(A)$, then A is bounded on $\{x \in \mathbb{R}^n : \|Bx\| \le 1\}$ and attains its maximum on that set.*

Proof. Let $(x_n)_{n \in \mathbb{N}}$ be a sequence in $\{x \in \mathbb{R}^n : \|Bx\| \le 1\}$. In accordance with Lemma 3, there exists a sequence $(z_n)_{n \in \mathbb{N}}$ in $\ker(B)$ such that $(x_n + z_n)_{n \in \mathbb{N}}$ is bounded. Since $A(x_n) = A(x_n + z_n)$ by hypothesis (recall that $\ker(B) \subseteq \ker(A)$), we conclude that A is bounded on $\{x \in \mathbb{R}^n : \|Bx\| \le 1\}$. Finally, let $(x_n)_{n \in \mathbb{N}}$ be a sequence in $\{x \in \mathbb{R}^n : \|Bx\| \le 1\}$ such that $\|Ax_n\| \to \max_{\|Bx\| \le 1} \|Ax\|$ as $n \to \infty$. Please note that $\|\overline{A}(x_n + \ker(B))\| = \|Ax_n\|$ for all $n \in \mathbb{N}$, so $(\overline{A}(x_n + \ker(B)))_{n \in \mathbb{N}}$ is bounded in $\mathbb{R}^m$ and so is $(\overline{A}(x_n + \ker(B)))_{n \in \mathbb{N}}$ in $\frac{\mathbb{R}^n}{\ker(B)}$. Fix $b_n \in \ker(B)$ such that $\|x_n + b_n\| < \|x_n + \ker(B)\| + \frac{1}{n}$ for all $n \in \mathbb{N}$. This means that $(x_n + b_n)_{n \in \mathbb{N}}$ is a bounded sequence in $\mathbb{R}^n$ so we can extract a convergent subsequence $(x_{n_k} + b_{n_k})_{k \in \mathbb{N}}$ to some $x_0 \in X$. At this stage, notice that $\|B(x_{n_k} + b_{n_k})\| = \|Bx_{n_k}\| \le 1$ for all $k \in \mathbb{N}$ and $(B(x_{n_k} + b_{n_k}))_{k \in \mathbb{N}}$ converges to Bx_0, so $\|Bx_0\| \le 1$. Note also that, since $\ker(B) \subseteq \ker(A)$, $(\|Ax_{n_k}\|)_{n \in \mathbb{N}}$ converges to $\|Ax_0\|$, which implies that

$$x_0 \in \arg \max_{\|Bx\| \le 1} \|Ax\|.$$

□

Theorem 8. *Let $A, B \in \mathbb{R}^{m \times n}$. The reformulated maxmin problem*

$$\begin{cases} \max \|Ax\| \\ \|Bx\| \le 1 \end{cases}$$

has a solution if and only if $\ker(B) \subseteq \ker(A)$.

Proof. If $\ker(B) \subseteq \ker(A)$, then we just need to call on Lemma 4. Conversely, if $\ker(B) \setminus \ker(A) \ne \varnothing$, then it suffices to consider the sequence $(nx_0)_{n \in \mathbb{N}}$ for $x_0 \in \ker(B) \setminus \ker(A)$, since $\|B(nx_0)\| = 0 \le 1$ for all $n \in \mathbb{N}$ and $\|A(nx_0)\| = n\|A(x_0)\| \to \infty$ as $n \to \infty$. □

4.5. Matrices on Quotient Spaces

Consider the maxmin

$$\begin{cases} \max \|T(x)\| \\ \|S(x)\| \le 1 \end{cases}$$

being X and Y Banach spaces and $T, S \in \mathcal{B}(X, Y)$ with $\ker(S) \subseteq \ker(T)$. Notice that if $(e_i)_{i \in I}$ is a Hamel basis of X, then $(e_i + \ker(S))_{i \in I}$ is a generator system of $\frac{X}{\ker(S)}$. By making use of the Zorn's Lemma, it can be shown that $(e_i + \ker(S))_{i \in I}$ contains a Hamel basis of $\frac{X}{\ker(S)}$. Observe that a subset C of $\frac{X}{\ker(S)}$ is linearly independent if and only if $S(C)$ is a linearly independent subset of Y.

In the finite dimensional case, we have

$$\overline{B} : \frac{\mathbb{R}^n}{\ker(B)} \to \mathbb{R}^m$$
$$x + \ker(B) \mapsto \overline{B}(x + \ker(B)) := Bx.$$

and

$$\overline{A} : \frac{\mathbb{R}^n}{\ker(B)} \to \mathbb{R}^m$$
$$x + \ker(B) \mapsto \overline{A}(x + \ker(B)) := Ax.$$

If $\{e_1, \ldots, e_n\}$ denotes the canonical basis of $\mathbb{R}^n$, then $\{e_1 + \ker(B), \ldots, e_n + \ker(B)\}$ is a generator system of $\frac{\mathbb{R}^n}{\ker(B)}$. This generator system contains a basis of $\frac{\mathbb{R}^n}{\ker(B)}$ so let $\{e_{j_1} + \ker(B), \ldots, e_{j_l} + \ker(B)\}$ be a basis of $\frac{\mathbb{R}^n}{\ker(B)}$. Please note that $\overline{A}(e_{j_k} + \ker(B)) = Ae_{j_k}$ and $\overline{B}(e_{j_k} + \ker(B)) = Be_{j_k}$ for every $k \in \{1, \ldots, l\}$. Therefore, the matrix associated with the linear map defined by $\overline{B}$ can be obtained from the matrix B by removing the columns corresponding to the indices $\{1, \ldots, n\} \setminus \{j_1, \ldots, j_l\}$, in other words, the matrix associated with $\overline{B}$ is $[Be_{j_1} | \cdots | Be_{j_l}]$. Similarly, the matrix associated with the linear map defined by $\overline{A}$ is $[Ae_{j_1} | \cdots | Ae_{j_l}]$. As we mentioned above, recall that a subset C of $\frac{\mathbb{R}^n}{\ker(B)}$ is linearly independent if and only if $B(C)$ is a linearly independent subset of $\mathbb{R}^m$. As a consequence, in order to obtain the basis $\{e_{j_1} + \ker(B), \ldots, e_{j_l} + \ker(B)\}$, it suffices to look at the rank of B and consider the columns of B that allow such rank, which automatically gives us the matrix associated with $\overline{B}$, that is, $[Be_{j_1} | \cdots | Be_{j_l}]$.

Finally, let

$$\pi : \mathbb{R}^n \to \frac{\mathbb{R}^n}{\ker(B)}$$
$$x \mapsto \pi(x) : x + \ker(B)$$

denote the quotient map. Let $l := \text{rank}(B) = \dim\left(\frac{\mathbb{R}^n}{\ker(B)}\right)$. If $x = (x_1, \ldots, x_l) \in \mathbb{R}^l$, then $\sum_{k=1}^{l} x_k (e_{j_k} + \ker(B)) \in \frac{\mathbb{R}^n}{\ker(B)}$. The vector $z \in \mathbb{R}^n$ defined by

$$z_p := \begin{cases} x_k & p = j_k \\ 0 & p \notin \{j_1, \ldots, j_l\} \end{cases}$$

verifies that

$$p(z) = \sum_{k=1}^{l} x_k (e_{j_k} + \ker(B)).$$

To simplify the notation, we can define the map

$$\alpha : \mathbb{R}^l \to \mathbb{R}^n$$
$$x \mapsto \alpha(x) := z$$

where z is the vector described right above.

5. Discussion

Here we compile all the results from the previous subsections and define the structure of the algorithm that solves the maxmin (3).

Let $A, B \in \mathbb{R}^{m \times n}$ with $\ker(B) \subseteq \ker(A)$. Then

$$\begin{cases} \max \|Ax\|_2 \\ \min \|Bx\|_2 \end{cases} \xrightarrow{\text{reform}} \begin{cases} \max \|Ax\|_2 \\ \|Bx\|_2 \leq 1 \end{cases}$$

Case 1: $\ker(B) = \{0\}$. B^+ denotes the Moore–Penrose inverse of B.

$$\begin{cases} \max \|Ax\|_2 \\ \|Bx\|_2 \leq 1 \end{cases} \xrightarrow{\text{supp. vec.}} \begin{cases} \max \|AB^+y\|_2 \\ \|y\|_2 \leq 1 \end{cases} \xrightarrow{\text{solution}} \begin{cases} y_0 \in \arg\max_{\|y\|_2 \leq 1} \|AB^+y\|_2 \\ \operatorname{rank}(B) = \operatorname{rank}([B|y_0]) \end{cases} \xrightarrow{\text{final sol.}} x_0 := B^+ y_0$$

Case 2: $\ker(B) \neq \{0\}$. $\overline{B} = [Be_{j_1}|\cdots|Be_{j_l}]$ where $\operatorname{rank}(B) = l = \operatorname{rank}(\overline{B})$ and $\overline{A} = [Ae_{j_1}|\cdots|Ae_{j_l}]$.

$$\begin{cases} \max \|Ax\|_2 \\ \|Bx\|_2 \leq 1 \end{cases} \xrightarrow{\text{case 1}} \begin{cases} \max \|\overline{A}y\|_2 \\ \|\overline{B}y\|_2 \leq 1 \end{cases} \xrightarrow{\text{solution}} y_0 \xrightarrow{\text{final sol.}} x_0 := \alpha(y_0)$$

In case a real-life problem is modeled like a maxmin involving more operators, we proceed as the following remark establishes in accordance with the preliminaries of this manuscript (reducing the number of multiobjective functions to avoid the lack of solutions):

Remark 1. Let $(T_n)_{n \in \mathbb{N}}$ and $(S_n)_{n \in \mathbb{N}}$ be sequences of continuous linear operators between Banach spaces X and Y. The maxmin

$$\begin{cases} \max \|T_n(x)\| \; n \in \mathbb{N} \\ \min \|S_n(x)\| \; n \in \mathbb{N} \end{cases} \tag{4}$$

can be reformulated as (recall the second typical reformulation)

$$\begin{cases} \max \sum_{n=1}^{\infty} \|T_n(x)\|^2 \\ \min \sum_{n=1}^{\infty} \|S_n(x)\|^2 \end{cases} \tag{5}$$

which can be transformed into a regular maxmin as in (1) by considering the operators

$$T: X \to \ell_2(Y)$$
$$x \mapsto T(x) := (T_n(x))_{n \in \mathbb{N}}$$

and

$$S: X \to \ell_2(Y)$$
$$x \mapsto S(x) := (S_n(x))_{n \in \mathbb{N}}$$

obtaining then

$$\begin{cases} \max \|T(x)\|^2 \\ \min \|S(x)\|^2 \end{cases}$$

which is equivalent to

$$\begin{cases} \max \|T(x)\| \\ \min \|S(x)\| \end{cases}$$

Observe that for the operators T and S to be well defined it is sufficient that $(\|T_n\|)_{n \in \mathbb{N}}$ and $(\|S_n\|)_{n \in \mathbb{N}}$ be in ℓ_2.

6. Materials and Methods

The initial methodology employed in this research work is the Mathematical Modelling of real-life problems. The subsequent methodology followed is given by the Axiomatic-Deductive Method framed in the First-Order Mathematical language. Inside this framework, we deal with the Category Theory (the main category involved is the Category of Banach spaces with the Bounded Operators). The final methodology used is the implementation of our mathematical results in the MATLAB programming language.

7. Conclusions

We finally enumerate the novelties provided in this work, which serve as conclusions for our research:

1. We prove that the original maxmin problem

$$\begin{cases} \max \|Ax\| \\ \min \|Bx\| \end{cases} \tag{6}$$

   has no solution (Theorem 2).

2. We then rewrite (6) as

$$\begin{cases} \max \|Ax\| \\ \|Bx\| \leq 1 \end{cases} \tag{7}$$

   which still models the real-life problem very accurately and has a solution if and only if $\ker(B) \subseteq \ker(A)$ (Theorem 8).

3. We provide an exact solution of (7) assuming $\ker(B) \subseteq \ker(A)$, not an heuristic method for approaching it. See Section 5.
4. A MATLAB code is provided for computing the solution to the maxmin problem. See Appendix C.
5. Our solution applies to design truly optimal minimum stored-energy TMS coils and to find more complex optimal geolocations involving statistical variables. See Appendixes A and B.
6. This article represents an interdisciplinary work involving pure abstract nontrivial proven theorems and programming codes that can be directly applied to different situations in the real world.

Author Contributions: Conceptualization, S.M.-P., F.J.G.-P., C.C.-S. and A.S.-A.; methodology, S.M.-P., F.J.G.-P., C.C.-S. and A.S.-A.; software, S.M.-P., F.J.G.-P., C.C.-S. and A.S.-A.; validation, S.M.-P., F.J.G.-P., C.C.-S. and A.S.-A.; formal analysis, S.M.-P., F.J.G.-P., C.C.-S. and A.S.-A.; investigation, S.M.-P., F.J.G.-P., C.C.-S. and A.S.-A.; resources, S.M.-P., F.J.G.-P., C.C.-S. and A.S.-A.; data curation, S.M.-P., F.J.G.-P., C.C.-S. and A.S.-A.; writing—original draft preparation, S.M.-P., F.J.G.-P., C.C.-S. and A.S.-A.; writing—review and editing, S.M.-P., F.J.G.-P., C.C.-S. and A.S.-A.; visualization, S.M.-P., F.J.G.-P., C.C.-S. and A.S.-A.; supervision, S.M.-P., F.J.G.-P., C.C.-S. and A.S.-A.; project administration, S.M.-P., F.J.G.-P., C.C.-S. and A.S.-A.; funding acquisition, S.M.-P., F.J.G.-P., C.C.-S. and A.S.-A. All authors have read and agreed to the published version of the manuscript.

Funding: This research was funded by the Research Grant PGC-101514-B-100 awarded by the Spanish Ministry of Science, Innovation and Universities and partially funded by FEDER.

Conflicts of Interest: The authors declare no conflict of interest. The funders had no role in the design of the study; in the collection, analyses, or interpretation of data; in the writing of the manuscript, or in the decision to publish the results.

Appendix A. Applications to Optimal TMS Coils

Appendix A.1. Introduction to TMS Coils

Transcranial Magnetic Stimulation (TMS) is a non-invasive technique to stimulate the brain. We refer the reader to [8,10,13–23] for a description on the development of TMS coils desing as an optimization problem.

An important safety issue in TMS is the minimization of the stimulation of non-target areas. Therefore, the development of TMS as a medical tool would be benefited with the design of TMS stimulators capable of inducing a maximum electric field in the region of interest, while minimizing the undesired stimulation in other prescribed regions.

Appendix A.2. Minimum Stored-Energy TMS Coil

In the following section, in order to illustrate an application of the theoretical model developed in this manuscript, we are going to tackle the design of a minimum stored-energy hemispherical TMS coil of radius 9 cm, constructed to stimulate only one cerebral hemisphere. To this end, the coil must produce an E-field which is both maximum in a spherical region of interest (ROI) and minimum in a second region (ROI2). Both volumes of interest are of 1 cm radius and formed by 400 points, where ROI is shifted by 5 cm in the positive z-direction and by 2 cm in the positive y-direction; and ROI2 is shifted by 5 cm in the positive z-direction and by 2 cm in the negative y-direction, as shown in Figure A1a. In Figure A1b a simple human head made of two compartments, scalp and brain, used to evaluate the performance of the designed stimulator is shown.

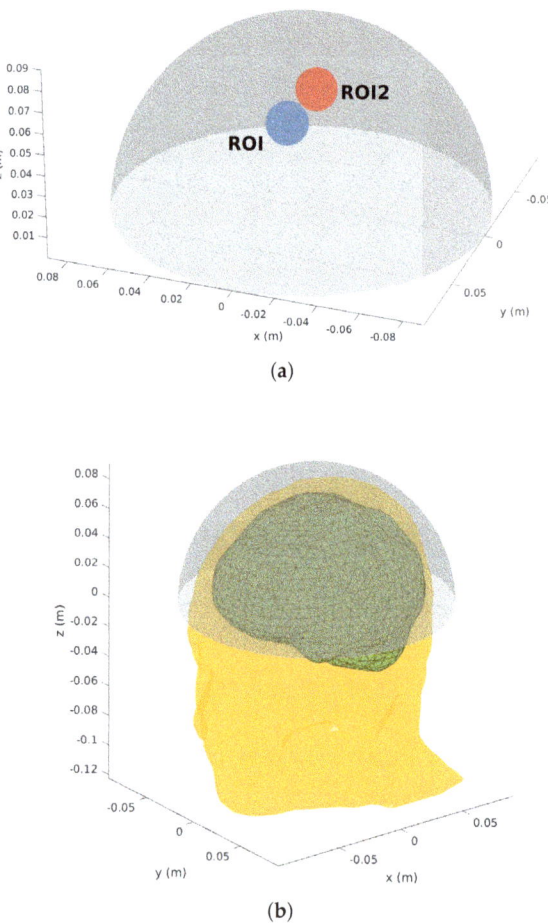

Figure A1. (a) Description of hemispherical surface where the optimal ψ must been found along with the spherical regions of interest ROI and ROI2 where the electric field must be maximized and minimized respectively. (b) Description of the two compartment scalp-brain model.

By using the formalism presented in [10] this TMS coil design problem can be posed as the following optimization problem:

$$\begin{cases} \max \|E_{x_1}\psi\|_2 \\ \min \|E_{x_2}\psi\|_2 \\ \min \psi^T L \psi \end{cases} \quad (A1)$$

where ψ is the stream function (the optimization variable), $M = 400$ are the number of points in the ROI and ROI2, $N = 2122$ the number of mesh nodes, $L \in \mathbb{R}^{N \times N}$ is the inductance matrix, and $E_{x_1} \in \mathbb{R}^{M \times N}$ and $E_{x_2} \in \mathbb{R}^{M \times N}$ are the E-field matrices in the prescribe x-direction.

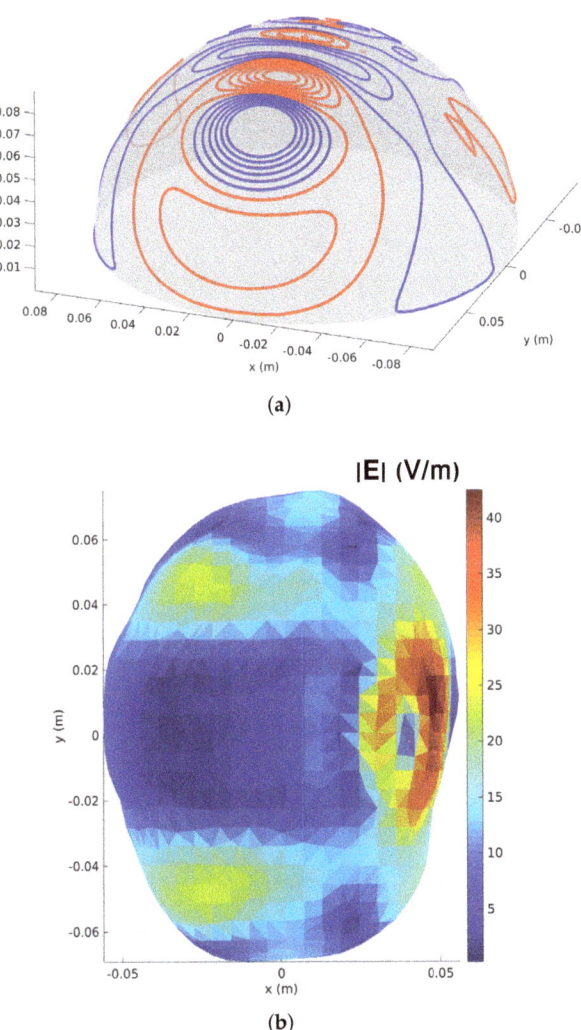

Figure A2. (a) Wirepaths with 18 turns of the TMS coil solution (red wires indicate reversed current flow with respect to blue). (b) E-field modulus induced at the surface of the brain by the designed TMS coil.

Figure A2a shows the coil solution of problem in Equation (A1) computed by using the theoretical model proposed in this manuscript (see Section 5 and Appendix A.3), and as expected, the wire arrangements is remarkably concentrated over the region of stimulation.

To evaluate the stimulation of the coil, we resort to the direct BEM [24], which permits the computation of the electric field induced by the coils in conducting systems. As can be seen in Figure A2b, the TMS coil fulfils the initial requirements of stimulating only one hemisphere of the brain (the one where ROI is found); whereas the electric field induced in the other cerebral hemisphere (where ROI2 can be found) is minimum.

Appendix A.3. Reformulation of Problem (A1) to Turn it into a Maxmin

Now it is time to reformulate the multiobjective optimization problem given in (A1), because it has no solution in virtue of Theorem 2. We will transform it into a maxmin problem as in (7) so that we can apply the theoretical model described in Section 5:

$$\begin{cases} \max \|E_{x_1}\psi\|_2 \\ \min \|E_{x_2}\psi\|_2 \\ \min \psi^T L \psi \end{cases}$$

Since raising to the square is a strictly increasing function on $[0, \infty)$, the previous problem is trivially equivalent to the following one:

$$\begin{cases} \max \|E_{x_1}\psi\|_2^2 \\ \min \|E_{x_2}\psi\|_2^2 \\ \min \psi^T L \psi \end{cases} \tag{A2}$$

Next, we apply Cholesky decomposition to L to obtain $L = C^T C$ so we have that $\psi^T L \psi = (C\psi)^T (C\psi) = \|C\psi\|_2^2$ so we obtain

$$\begin{cases} \max \|E_{x_1}\psi\|_2^2 \\ \min \|E_{x_2}\psi\|_2^2 \\ \min \|C\psi\|_2^2 \end{cases} \tag{A3}$$

Since C is an invertible square matrix, $\arg\min \|C\psi\|_2^2 = \{0\}$ so the previous multiobjective optimization problem has no solution. Therefore it must be reformulated. We call then on Remark 1 to obtain:

$$\begin{cases} \max \|E_{x_1}\psi\|_2^2 \\ \min \|E_{x_2}\psi\|_2^2 + \|C\psi\|_2^2 \end{cases} \tag{A4}$$

which in essence is

$$\begin{cases} \max \|E_{x_1}\psi\|_2 \\ \min \|D\psi\|_2 \end{cases} \tag{A5}$$

where $D := \begin{pmatrix} E_{x_2} \\ C \end{pmatrix}$. The matrix D in this specific case has null kernel. In accordance with the previous sections, Problem (A5) is remodeled as

$$\begin{cases} \max \|E_{x_1}\psi\|_2 \\ \|D\psi\|_2 \leq 1 \end{cases} \tag{A6}$$

Finally, we can refer to Section 5 to solve the latter problem.

Appendix B. Applications to Optimal Geolocation

Several studies involving optimal geolocation [25], multivariate statistics [26,27] and multiobjective problems [28–30] were carried out recently. To show another application of maxmin multiobjective problems, we consider in this work the best situation of a tourism rural inn considering several measured climate variables. Locations with low highest temperature m_1, radiation m_2 and evapotranspiration m_3 in summer time and high values in winter time are sites with climatic characteristics desirable for potential visitors. To solve this problem, we choose 11 locations in the Andalusian coastline and 2 in the inner, near the mountains. We have collected the data from the official *Andalusian government* webpage [31] evaluating the mean values of these variables on the last 5 years 2013–2019. The referred months of the study were January and July.

Table A1. Mean values of high temperature (T) in Celsius Degrees, radiation (R) in MJ/m^2, and evapotranspiration (E) in mm/day, measures in January (winter time) and July (summer time) between 2013 and 2018.

	T-Winter	R-Winter	E-Winter	T-Summer	R-Summer	E-Summer
Sanlúcar	15.959	9.572	1.520	30.086	27.758	6.103
Moguer	16.698	9.272	0.925	30.424	27.751	5.222
Lepe	16.659	9.503	1.242	30.610	28.297	6.836
Conil	16.322	9.940	1.331	28.913	26.669	5.596
El Puerto	16.504	9.767	1.625	31.052	28.216	6.829
Estepona	16.908	10.194	1.773	31.233	27.298	6.246
Málaga	17.663	9.968	1.606	32.358	27.528	6.378
Vélez	18.204	9.819	1.905	31.912	26.534	5.911
Almuñécar	17.733	10.247	1.404	29.684	25.370	4.952
Adra	17.784	10.198	1.637	28.929	26.463	5.143
Almería	17.468	10.068	1.561	30.342	27.335	5.793
Aroche	16.477	9.797	1.434	34.616	27.806	6.270
Córdoba	14.871	8.952	1.149	36.375	28.503	7.615
Baza	13.386	8.303	3.054	35.754	27.824	1.673
Bélmez	13.150	8.216	1.215	35.272	28.478	7.400
S. Yeguas	13.656	9.155	1.247	33.660	28.727	7.825

To find the optimal location, let us evaluate the site where the variables mean values are maximum in January and minimum in July. Here we have a typical multiobjective problem with two data matrices that can be formulated as follows:

$$\begin{cases} \max \|Ax\|_2 \\ \min \|Bx\|_2 \\ \min \|x\|_2 \end{cases} \quad (A7)$$

where A and B are real 16×3 matrices with the values of the three variables (m_1, m_2, m_3) taking into account (highest temperature, radiation and evapotranspiration) in January and July respectively. To avoid unit effects, we standarized the variables ($\mu = 0$ and $\sigma = 1$). The vector x is the solution of the multiobjective problem.

Since (A7) lacks any solution in view of Theorem 2, we reformulate it as we showed in Remark 1 by the following:

$$\begin{cases} \max \|Ax\|_2 \\ \min \|Dx\|_2 \end{cases} \quad (A8)$$

with matrix $D := \begin{pmatrix} B \\ I_n \end{pmatrix}$, where I_n is the identity matrix with $n = 3$. Notice that it also verifies that $\ker(D) = \{0\}$. Observe that, according to the previous sections, (A8) can be remodeled into

$$\begin{cases} \max \|Ax\|_2 \\ \|Dx\|_2 \leq 1 \end{cases} \quad (A9)$$

and solved accordingly.

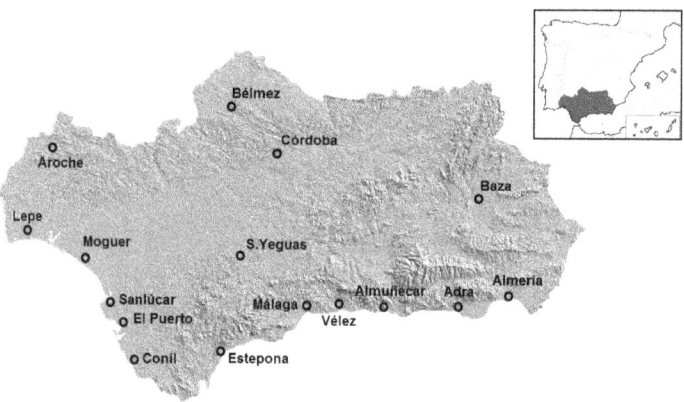

Figure A3. Geographic distribution of the sites considered in the study. 11 places are in the coastline of the region and 5 in the inner.

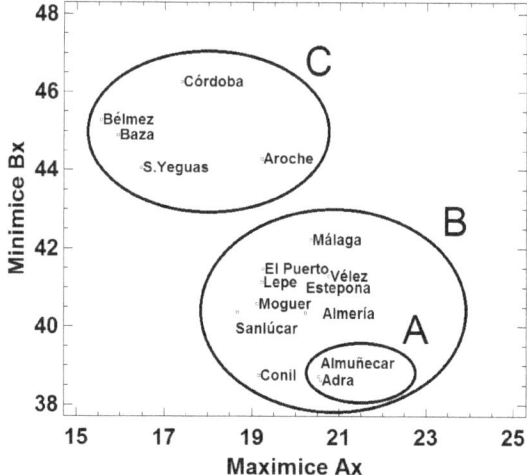

Figure A4. Locations considering Ax and Bx axes. Group named A represents the best places for the tourism rural inn, near Costa Tropical (Granada province). Sites on B are also in the coastline of the region. Sites on C are the worst locations considering the multiobjective problem, they are situated inside the region.

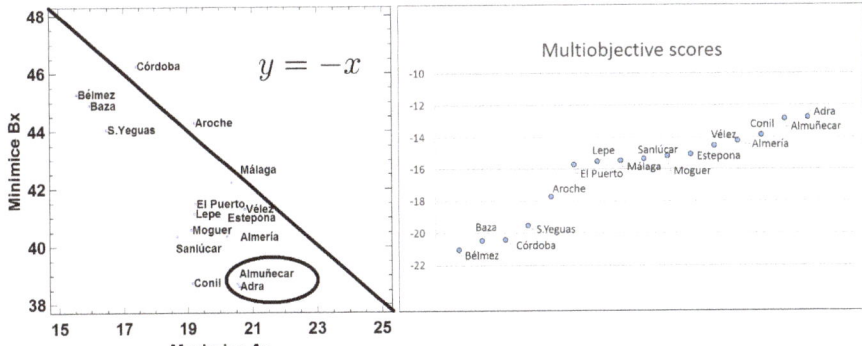

Figure A5. (**left**) Sites considering Ax and Bx and the function $y = -x$. The places with high values of Ax (max) and low values of Bx (min) are the best locations for the solution of the multiobjective problem (round). (**right**) Multiobjective scores values obtained for each site projecting the point in the function $y = -x$. High values of this score indicate better places to locate the tourism rural inn.

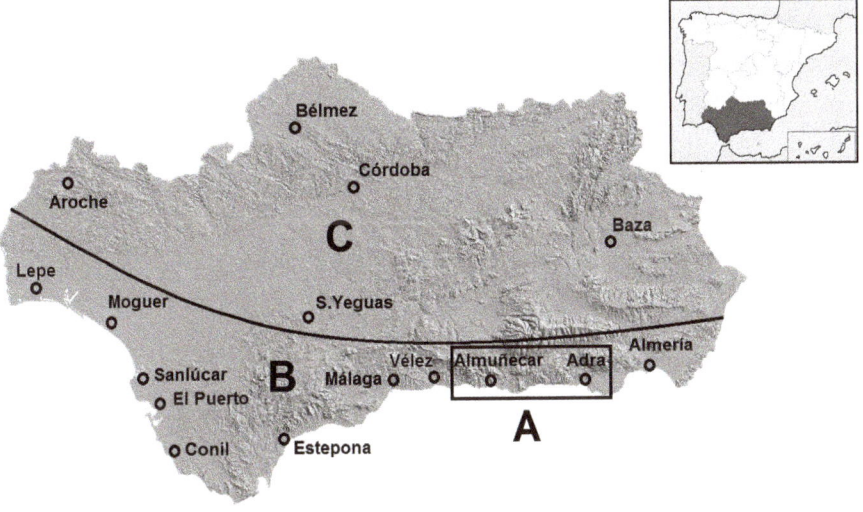

Figure A6. Distribution of the three areas described in Figure A4. A and B areas are in the coastline and C in the inner.

The solution of (A9) allow us to draw the sites with a 2D plot considering the X axe as Ax and the Y axe as Bx. We observe that better places have high values of Ax and low values of Bx. Hence, we can sort the sites in order to achieve the objectives in a similar way as factorial analysis works (two factors, the maximum and the minimum, instead of m variables).

Appendix C. Algorithms

To solve the real problems posed in this work, the algorithms were developed in MATLAB. As pointed out in Section 5, our method relies on finding the generalized supporting vectors. Thus, we refer the reader to [8] (Appendix A.1) for the MATLAB code "sol_1.m" to compute a basis of generalized supporting vectors of a finite number of matrices $A_1, \ldots, A_k$, in other words, a solution of Problem (A10), which was originally posed and solved in [7]:

$$\begin{cases} \max \sum_{i=1}^{k} \|A_i x\|_2^2 \\ \|x\|_2 = 1 \end{cases} \tag{A10}$$

The solution of the previous problem (see [7] (Theorem 3.3)) is given by

$$\max_{\|x\|_2=1} \sum_{i=1}^{k} \|A_i x\|_2^2 = \lambda_{\max} \left(\sum_{i=1}^{k} A_i^T A_i \right)$$

and

$$\arg\max_{\|x\|_2=1} \sum_{i=1}^{k} \|A_i x\|_2^2 = V\left(\lambda_{\max} \left(\sum_{i=1}^{k} A_i^T A_i \right) \right) \cap S_{\ell_2^n}$$

where $\lambda_{\max}$ denotes the greatest eigenvalue and V denotes the associated eigenvector space. We refer the reader to [8] (Theorem 4.2) for a generalization of [7] (Theorem 3.3) to a infinite number of operators on an infinite dimensional Hilbert space.

As we pointed out in Theorem 8, the solution of the problem

$$\begin{cases} \max \|Ax\| \\ \|Bx\| \leq 1 \end{cases}$$

exists if and only if $\ker(B) \subseteq \ker(A)$. Here is a simple code to check this.

```
function p=existence_sol(A,B)
%%%
%%% This function checks the existence of the solution of the
%%% problem
%%%
%%% max ||Ax||
%%% ||Bx||<=1
%%%
%%%%%%%%%%%%%%%%%%%%%%%%%%%%%%%%
%%%
%%% INPUT:
%%%
%%% A, B - the matrices involved in the problem
%%%
%%%%%%%%%%%%%%%%%%%%%%%%%%%%%
%%%
%%% OUTPUT:
%%%
%%% p - true if the problem has solution or false on the contrary
%%%
%%%%%%%%%%%%%%%%%%%%%%%%%%%%%
KerB = null(B);
dimKerB = size(KerB,2);
KerA = null(A);
dimKerA = size(KerA,2);
if (dimKerB<=dimKerA) & (rank([KerB KerA])==dimKerA)
    p = true;
else
```

```
        p = false;
    end
end
```

Now we present the code to solve the first case of the previous maxmin problem, that is, the case where $\ker(B) = \{0\}$. We refer the reader to Section 5 on which this code is based.

```
function x = case_1(A, B)
    %%%%
    %%%% This function computes the solution of the problem
    %%%%
    %%%% max ||Ax||_2
    %%%% ||Bx||_2<=1
    %%%%
    %%%% in the case KerB={0}.
    %%%%%%%%%%%%%%%%%%%%%%%%%%%%
    %%%%
    %%%% INPUT:
    %%%%
    %%%% A, B - the matrices involved in the problem
    %%%%
    %%%%%%%%%%%%%%%%%%%%%%%%%%%%
    %%%%
    %%%% OUTPUT:
    %%%%
    %%%% x - basis of unit eigenvectors associated to lambda_max
    %%%%
    %%%%%%%%%%%%%%%%%%%%%%%%%%%%
    %%%%
    KerB = null(B);
    dimKerB = size(KerB,2);
    if (dimKerB ~= 0)
        display('KerB~={0}')
        x=[];
    else % KerB={0}
        M = A*pinv(B);            % M = A*B^+
                                  % B^+ is the pseudoinverse matrix
        [lambda_max, y] = sol_1({M}); % where sol_1 is the algorithm in [5, Appendix A.1]
        [nrows_y ncols_y] = size(y);
        r_B = rank(B);
        counter = 0;
        for i=1:ncols_y
            r = rank([B y(:,i)]);
            if (abs(r_B - r)<1e-12)   % Here we check if rank(B) = rank ([B y0]).
                                      % A tolerance of 1e-12 is needed in
                                      % order to compare these two ranks.
                counter = counter +1;
                y0(:,counter) = y(:,i);
            end
        end
        x = pinv(B)*y0;           % This is a basis of solutions of our problem
end
```

Next, we can compute the global solution of the maxmin problem by means of the following code. Again, we refer the reader to Section 5 on which this code is based.

```matlab
function x = sol_2(A, B)
%%%%
%%%% This function computes the solution of the problem
%%%%
%%%% max ||Ax||_2
%%%% ||Bx||_2<=1
%%%%
%%%%%%%%%%%%%%%%%%%%%%%%%%%%%%%
%%%%
%%%% INPUT:
%%%%
%%%% A, B - the matrices involved in the problem
%%%%
%%%%%%%%%%%%%%%%%%%%%%%%%%%%%%%
%%%%
%%%% OUTPUT:
%%%%
%%%% x - Supporting vector which is the solution of the problem
%%%%
%%%%%%%%%%%%%%%%%%%%%%%%%%%%%%%
%%%%
p=existence_sol(A,B);
if p==true
    n = size(B,2);
    KerB = null(B);
    dimKerB = size(KerB,2);
    if (dimKerB == 0)            % KerB = {0} This is the case 1
        x = case_1(A,B);         % x is the solution of our problem
    else % KerB~={0}
        [Br indices] = colsindep(B); %%% First we extract the
                                     %%% independent columns in B
        Ar = A(indices);     %%% We extract the same columns of A
            %%% Now, Ker(Br)={0} so this is the case 1 treated above:
        xr = case_1(Ar,Br);
        [nrows_xr,ncols_xr] = size(xr);
        %%% Now we compute the matrix solutions x of the problem
        counter = 0;
        for j = 1:ncols_xr
           for i=1:n
               if ismember(i,indices)==1 %%% i is an index of the ones
                                         %%% defined above
                   counter = counter + 1;
                   x(i,j) = xr(counter,j);
               else
                   x(i,j) = 0;
               end
           end
        end
    end

else
    display('This problem has no solution');
    x=[];
end
```

Notice that we use the case_1 function described above and a new function named colsindep. We include the code to implement this new function below.

```matlab
function [Dcolsind, indices]=colsindep(D)
%%%%
%%%% This function extracts r = rank(D) independent columns of the
%%%% matrix D and the indices of the columns in D which are independent
%%%%
%%%%%%%%%%%%%%%%%%%%%%%%%%%%%%%
%%%%
%%%% INPUT:
%%%%
%%%% D - a matrix with rank r
%%%%
%%%%%%%%%%%%%%%%%%%%%%%%%%%%%%%
%%%%
%%%% OUTPUT:
%%%%
%%%% Dcolsind - r independent columns in D
%%%% indices - the indices of independent columns extracted from D
%%%%%%%%%%%%%%%%%%%%%%%%%%%%%%%
    r=rank(D);              %%% Compute the rank
    [Q R p]=qr(D,0);        %%% p is a permutation vector such that A(:,p)=Q*R
    indices=sort(p(1:r));   %%% The first r elements in p are the indices of the
                            %%% columns linearly independent in D
    Dcolsind=D(:,indices);  %%% Extract these columns
end
```

The MATLAB code to compute the solution of the TMS coil problem (A6):

$$\begin{cases} \max \|E_{x_1}\psi\|_2 \\ \|D\psi\|_2 \leq 1 \end{cases}$$

with the matrix $D := \begin{pmatrix} E_{x_2} \\ C \end{pmatrix}$, where C is the Cholesky matrix of L, and in this case it verifies that $\ker(D) = \{0\}$. Recall that (A6) comes from (A1):

$$\begin{cases} \max \|E_{x_1}\psi\|_2 \\ \min \|E_{x_2}\psi\|_2 \\ \min \psi^T L \psi \end{cases}$$

```matlab
function psi = sol2_psi(Ex1, Ex2, L)

    C = chol(L);            % Cholesky's decomposition of matrix L = C' * C

    A = Ex1;
    B = [Ex2;C];

    psi = case_1(A,B);      % We apply the algorithm to obtain the solutions
end
```

Finally, we provide the code to compute the solution of the optimal geolocation problem (A9):

$$\begin{cases} \max \|Ax\|_2 \\ \|Dx\|_2 \leq 1 \end{cases}$$

with matrix $D := \begin{pmatrix} B \\ I_3 \end{pmatrix}$. Notice that it also verifies that $\ker(D) = \{0\}$ and A and B are composed by standardized variables. Recall that (A9) comes from (A7):

$$\begin{cases} \max \|Ax\|_2 \\ \min \|Bx\|_2 \\ \min \|x\|_2 \end{cases}$$

```
function x = sol_2_geoloc(A, B)

    [rows,cols] = size(A);
    D = [B; eye(size(cols))];

    x = case_1(A,D);         % We apply the algorithm to obtain the solutions
end
```

References

1. Huang, N.; Ma, C.F. Modified conjugate gradient method for obtaining the minimum-norm solution of the generalized coupled Sylvester-conjugate matrix equations. *Appl. Math. Model.* **2016**, *40*, 1260–1275, doi:10.1016/j.apm.2015.07.017. [CrossRef]
2. Yassin, B.; Lahcen, A.; Zeriab, E.S.M. Hybrid optimization procedure applied to optimal location finding for piezoelectric actuators and sensors for active vibration control. *Appl. Math. Model.* **2018**, *62*, 701–716, doi:10.1016/j.apm.2018.06.017. [CrossRef]
3. Bishop, E.; Phelps, R.R. A proof that every Banach space is subreflexive. *Bull. Am. Math. Soc.* **1961**, *67*, 97–98, doi:10.1090/S0002-9904-1961-10514-4. [CrossRef]
4. Bishop, E.; Phelps, R.R. The support functionals of a convex set. In *Proceedings of Symposia in Pure Mathematics*; American Mathematical Society: Providence, RI, USA, 1963; Volume VII, pp. 27–35.
5. Lindenstrauss, J. On operators which attain their norm. *Israel J. Math.* **1963**, *1*, 139–148, doi:10.1007/BF02759700. [CrossRef]
6. James, R.C. Characterizations of reflexivity. *Stud. Math.* **1964**, *23*, 205–216, doi:10.4064/sm-23-3-205-216. [CrossRef]
7. Cobos-Sánchez, C.; García-Pacheco, F.J.; Moreno-Pulido, S.; Sáez-Martínez, S. Supporting vectors of continuous linear operators. *Ann. Funct. Anal.* **2017**, *8*, 520–530, doi:10.1215/20088752-2017-0016. [CrossRef]
8. Garcia-Pacheco, F.J.; Cobos-Sanchez, C.; Moreno-Pulido, S.; Sanchez-Alzola, A. Exact solutions to $\max_{\|x\|=1} \sum_{i=1}^{\infty} \|T_i(x)\|^2$ with applications to Physics, Bioengineering and Statistics. *Commun. Nonlinear Sci. Numer. Simul.* **2020**, *82*, 105054, doi:10.1016/j.cnsns.2019.105054. [CrossRef]
9. García-Pacheco, F.J.; Naranjo-Guerra, E. Supporting vectors of continuous linear projections. *Int. J. Funct. Anal. Oper. Theory Appl.* **2017**, *9*, 85–95. [CrossRef]
10. Cobos Sánchez, C.; Garcia-Pacheco, F.J.; Guerrero Rodriguez, J.M.; Hill, J.R. An inverse boundary element method computational framework for designing optimal TMS coils. *Eng. Anal. Bound. Elem.* **2018**, *88*, 156–169, doi:10.1016/j.enganabound.2017.11.002. [CrossRef]
11. Bohnenblust, F. A characterization of complex Hilbert spaces. *Portugal. Math.* **1942**, *3*, 103–109.
12. Kakutani, S. Some characterizations of Euclidean space. *Jpn. J. Math.* **1939**, *16*, 93–97, doi:10.4099/jjm1924.16.0_93. [CrossRef]
13. Sánchez, C.C.; Rodriguez, J.M.G.; Olozábal, Á.Q.; Blanco-Navarro, D. Novel TMS coils designed using an inverse boundary element method. *Phys. Med. Biol.* **2016**, *62*, 73–90, doi:10.1088/1361-6560/62/1/73. [CrossRef]

14. Marin, L.; Power, H.; Bowtell, R.W.; Cobos Sanchez, C.; Becker, A.A.; Glover, P.; Jones, A. Boundary element method for an inverse problem in magnetic resonance imaging gradient coils. *Comput. Model. Eng. Sci.* **2008**, *23*, 149–173.
15. Marin, L.; Power, H.; Bowtell, R.W.; Cobos Sanchez, C.; Becker, A.A.; Glover, P.; Jones, I.A. Numerical solution of an inverse problem in magnetic resonance imaging using a regularized higher-order boundary element method. In *Boundary Elements and Other Mesh Reduction Methods XXIX*; WIT Press: Southampton, UK, 2007; Volume 44, pp. 323–332, doi:10.2495/BE070311. [CrossRef]
16. Wassermann, E.; Epstein, C.; Ziemann, U.; Walsh, V.; Paus, T.; Lisanby, S. *Oxford Handbook of Transcranial Stimulation (Oxford Handbooks)*, 1st ed.; Oxford University Press: New York, NY, USA, 2008.
17. Romei, V.; Murray, M.M.; Merabet, L.B.; Thut, G. Occipital Transcranial Magnetic Stimulation Has Opposing Effects on Visual and Auditory Stimulus Detection: Implications for Multisensory Interactions. *J. Neurosci.* **2007**, *27*, 11465–11472, doi:10.1523/JNEUROSCI.2827-07.2007. [CrossRef]
18. Koponen, L.M.; Nieminen, J.O.; Ilmoniemi, R.J. Minimum-energy Coils for Transcranial Magnetic Stimulation: Application to Focal Stimulation. *Brain Stimul.* **2015**, *8*, 124–134, doi:10.1016/j.brs.2014.10.002. [CrossRef]
19. Koponen, L.M.; Nieminen, J.O.; Mutanen, T.P.; Stenroos, M.; Ilmoniemi, R.J. Coil optimisation for transcranial magnetic stimulation in realistic head geometry. *Brain Stimul.* **2017**, *10*, 795–805, doi:10.1016/j.brs.2017.04.001. [CrossRef]
20. Gomez, L.J.; Goetz, S.M.; Peterchev, A.V. Design of transcranial magnetic stimulation coils with optimal trade-off between depth, focality, and energy. *J. Neural Eng.* **2018**, *15*, 046033, doi:10.1088/1741-2552/aac967. [CrossRef]
21. Wang, B.; Shen, M.R.; Deng, Z.D.; Smith, J.E.; Tharayil, J.J.; Gurrey, C.J.; Gomez, L.J.; Peterchev, A.V. Redesigning existing transcranial magnetic stimulation coils to reduce energy: application to low field magnetic stimulation. *J. Neural Eng.* **2018**, *15*, 036022, doi:10.1088/1741-2552/aaa505. [CrossRef]
22. Grandy, W.T. Time Evolution in Macroscopic Systems. I. Equations of Motion. *Found. Phys.* **2004**, *34*, 1–20, doi:10.1023/B:FOOP.0000012007.06843.ed. [CrossRef]
23. Sakurai, J.J. *Modern Quantum Mechanics*; Addison-Wesley Publishing Company: Reading, MA, USA, 1993.
24. Sanchez, C.C.; Bowtell, R.W.; Power, H.; Glover, P.; Marin, L.; Becker, A.A.; Jones, A. Forward electric field calculation using BEM for time-varying magnetic field gradients and motion in strong static fields. *Eng. Anal. Bound. Elem.* **2009**, *33*, 1074–1088, doi:10.1016/j.enganabound.2009.02.006. [CrossRef]
25. Jäntschi, L.; Bálint, D.; Bolboaca, S. Multiple Linear Regressions by Maximizing the Likelihood under Assumption of Generalized Gauss-Laplace Distribution of the Error. *Comput. Math. Methods Med.* **2016**, *2016*, 1–8, doi:10.1155/2016/8578156. [CrossRef] [PubMed]
26. Gil-García, I.C.; García-Cascales, M.S.; Fernández-Guillamón, A.; Molina-García, A. Categorization and Analysis of Relevant Factors for Optimal Locations in Onshore and Offshore Wind Power Plants: A Taxonomic Review. *J. Mar. Sci. Eng.* **2019**, *7*, 391, doi:10.3390/jmse7110391. [CrossRef]
27. Pérez Morales, A.; Castillo, F.; Pardo-Zaragoza, P. Vulnerability of Transport Networks to Multi-Scenario Flooding and Optimum Location of Emergency Management Centers. *Water* **2019**, *11*, 1197, doi:10.3390/w11061197. [CrossRef]
28. Choi, J.W.; Kim, M.K. Multi-Objective Optimization of Voltage-Stability Based on Congestion Management for Integrating Wind Power into the Electricity Market. *Appl. Sci.* **2017**, *7*, 573, doi:10.3390/app7060573. [CrossRef]
29. Zavala, G.R.; García-Nieto, J.; Nebro, A.J. Qom—A New Hydrologic Prediction Model Enhanced with Multi-Objective Optimization. *Appl. Sci.* **2019**, *10*, 251, doi:10.3390/app10010251. [CrossRef]
30. Susowake, Y.; Masrur, H.; Yabiku, T.; Senjyu, T.; Motin Howlader, A.; Abdel-Akher, M.; Hemeida, A.M. A Multi-Objective Optimization Approach towards a Proposed Smart Apartment with Demand-Response in Japan. *Energies* **2019**, *13*, 127, doi:10.3390/en13010127. [CrossRef]
31. ESTACIONES AGROCLIMÁTICAS. Available online: https://www.juntadeandalucia.es/agriculturaypesca/ifapa/ria/servlet/FrontController (accessed on 18 September 2019).

© 2020 by the authors. Licensee MDPI, Basel, Switzerland. This article is an open access article distributed under the terms and conditions of the Creative Commons Attribution (CC BY) license (http://creativecommons.org/licenses/by/4.0/).

Article

On q-Quasi-Newton's Method for Unconstrained Multiobjective Optimization Problems

Kin Keung Lai [1,*,†], Shashi Kant Mishra [2,†] and Bhagwat Ram [3,†]

1 College of Economics, Shenzhen University, Shenzhen 518060, China
2 Department of Mathematics, Institute of Science, Banaras Hindu University, Varanasi 221005, India; shashikant.mishra@bhu.ac.in
3 DST-Centre for Interdisciplinary Mathematical Sciences, Institute of Science, Banaras Hindu University, Varanasi 221005, India; bhagwat.ram2@bhu.ac.in
* Correspondence: mskklai@outlook.com
† These authors contributed equally to this work.

Received: 01 April 2020; Accepted: 13 April 2020; Published: 17 April 2020

Abstract: A parameter-free optimization technique is applied in Quasi-Newton's method for solving unconstrained multiobjective optimization problems. The components of the Hessian matrix are constructed using q-derivative, which is positive definite at every iteration. The step-length is computed by an Armijo-like rule which is responsible to escape the point from local minimum to global minimum at every iteration due to q-derivative. Further, the rate of convergence is proved as a superlinear in a local neighborhood of a minimum point based on q-derivative. Finally, the numerical experiments show better performance.

Keywords: multiobjective programming; methods of quasi-Newton type; Pareto optimality; q-calculus; rate of convergence

MSC: 90C29; 90C53; 58E17; 05A30; 41A25

1. Introduction

Multiobjective optimization is the method of optimizing two or more real valued objective functions at the same time. There is no ideal minimizer to minimize all objective functions at once, thus the optimality concept is replaced by the idea of Pareto optimality/efficiency. A point is called Pareto optimal or efficient if there does not exist an alternative point with the equivalent or smaller objective function values, such that there is a decrease in at least one objective function value. In many applications such as engineering [1,2], economic theory [3], management science [4], machine learning [5,6], and space exploration [7], etc., several multiobjective optimization techniques are used to make the desired decision. One of the basic approaches is the weighting method [8], where a single objective optimization problem is created by the weighting of several objective functions. Another approach is the ϵ-constraint method [9], where we minimize only the chosen objective function and keep other objectives as constraints. Some multiobjective algorithms require a lexicographic method, where all objective functions are optimized in their order of priority [10,11]. First, the most preferred function is optimized, then that objective function is transformed into a constraint and a second priority objective function is optimized. This approach is repeated until the last objective function is optimized. The user needs to choose the sequence of objectives. Two distinct lexicographic optimizations with distinct sequences of objective functions do not produce the same solution. The disadvantages of such approaches are the choice of weights, constraints, and importance of the functions, respectively, which are not known in advance and they have to be specified from the beginning. Some other techniques [12–14] that do not need any prior information are developed for solving unconstrained multiobjective optimization problems (UMOP) with at most linear convergence rate. Other methods like heuristic approaches or evolutionary approaches [15] provide an approximate Pareto front but do not guarantee the convergence property.

Newton's method [16] that solves the single-objective optimization problems is extended for solving (UMOP), which is based on an a priori parameter-free optimization method [17]. In this case, the objective functions are twice continuously differentiable, no other parameter or ordering of the functions is needed, and each objective function is replaced with a quadratic model. The rate of convergence is observed as superlinear, and it is quadratic if the second-order derivative is Lipschitz continuous. Newton's method is also studied under the assumptions of Banach and Hilbert spaces for finding the efficient solutions of (UMOP) [18]. A new type of Quasi-Newton algorithm is developed to solve the nonsmooth multiobjective optimization problems, where the directional derivative of every objective function exists [19].

A necessary condition for finding the vector critical point of (UMOP) is introduced in the steepest descent algorithm [12], where neither weighting factors nor ordering information for the different objective functions are assumed to be known. The relationship between critical points and efficient points is discussed in [17]. If the domain of (UMOP) is a convex set and the objective functions are convex component-wise then every critical point is the weak efficient point, and if the objective functions are strictly convex component-wise, then every critical point is the efficient point. The new classes of vector invex and pseudoinvex functions for (UMOP) are also characterized in terms of critical points and (weak) efficient points [20] by using Fritz John (FJ) optimality conditions and Karush–Kuhn–Tucker (KKT) conditions. Our focus is on Newton's direction for a standard scalar optimization problem which is implicitly induced by weighting the several objective functions. The weighting values are a priori unknown and non-negative KKT multipliers, that is, they are not required to fix in advance. Every new point generated by the Newton algorithm [17] initiates such weights in the form of KKT multipliers.

Quantum calculus or q-calculus is also called calculus without limits. The q-analogues of mathematical objects can be again recaptured as $q \to 1$. The history of quantum calculus can be traced back to Euler (1707–1783), who first proposed the quantum q in Newton's infinite series. In recent years, many researchers have shown considerable interest in examining and exploring the quantum calculus. Therefore, it emerges as an interdisciplinary subject. Of course, the quantum analysis is very useful in numerous fields such as in signal processing [21], operator theory [22], fractional integral and derivatives [23], integral inequalities [24], variational calculus [25], transform calculus [26], sampling theory [27], etc. The quantum calculus is seen as the bridge between mathematics and physics. To study some recent developments in quantum calculus, interested researches should refer to [28–31].

The q-calculus was first studied in the area of optimization [32], where the q-gradient is used in steepest descent method to optimize objective functions. Further, global optimum was searched using q-steepest descent method and q-conjugate gradient method where a descent scheme is presented using q-calculus with the stochastic approach which does not focus on the order of convergence of the scheme [33]. The q-calculus is applied in Newton's method to solve unconstrained single objective optimization [34]. Further, this idea is extended to solve (UMOP) within the context of the q-calculus [35].

In this paper, we present the q-calculus in Quasi-Newton's method for solving (UMOP). We approximate the second q-derivative matrices instead of evaluating them. Using q-calculus, we present the convergence rate is superlinear.

The rest of this paper is organized as follows. Section 2 recalls the problem, notation, and preliminaries. Section 3 derives a q-Quasi-Newton direction search method solved by (KKT) conditions. Section 4 establishes the algorithms for convergence analysis. The numerical results are given in Section 5 and the conclusion is in the last section.

2. Preliminaries

Denote $\mathbb{R}$ as the set of real numbers, $\mathbb{N}$ as the set of positive integers, and $\mathbb{R}_+$ or $(\mathbb{R}_-)$ as the set of strictly positive or (negative) real numbers. If a function is continuous on any interval excluding zero, then the function is called continuous q-differentiable. For a function $f : \mathbb{R} \to \mathbb{R}$, the q-derivative of f [36] denoted as $D_{q,x}f$, is given as

$$D_{q,x}f(x) = \begin{cases} \frac{f(x)-f(qx)}{(1-q)x}, & x \neq 0, q \neq 1 \\ f'(x), & x = 0. \end{cases} \quad (1)$$

Suppose $f : \mathbb{R}^n \to \mathbb{R}$, whose partial derivatives exist. For $x \in \mathbb{R}^n$, consider an operator $\epsilon_{q,i}$ on f as

$$(\epsilon_{q,i})f(x) = f(x_1, x_2, \ldots, qx_i, x_{i+1}, \ldots, x_n). \quad (2)$$

The q-partial derivative of f at x with respect to x_i, indicated by $D_{q,x_i}f$, is [23]:

$$D_{q,x_i}f(x) = \begin{cases} \frac{f(x)-(\epsilon_{q,i}f)(x)}{(1-q)x_i}, & x_i \neq 0, q \neq 1, \\ \frac{\partial f}{\partial x_i}, & x_i = 0. \end{cases} \quad (3)$$

We are interested to solve the following (UMOP):

$$\text{minimize} \quad F(x) \quad (4)$$
$$\text{subject to} \quad x \in X,$$

where $X \subseteq \mathbb{R}^n$ is a feasible region and $F : X \to \mathbb{R}^m$. Note that the function $F = (f_1, f_2, \ldots, f_m)$ is a vector function whose components are real valued functions such as $f_j : X \to \mathbb{R}$, where $j = 1, \ldots, m$. In general, n and m are independent. For $x, y \in \mathbb{R}^n$, we present the vector inequalities as:

$$x = y \iff x_i = y_i; \forall i = 1, \ldots, n,$$
$$x \geqq y \iff x_i \geq y_i \; \forall i = 1, \ldots, n,$$
$$x \geq y \iff x_i \geq y_i \text{ and } x \neq y,$$
$$x > y \iff x_i > y_i \; \forall i = 1, \ldots, n.$$

A point $x^* \in X$ is called Pareto optimal point such that there is no any point $x \in X$, for which $F(x) \leq F(x^*)$, and $F(x) \neq F(x^*)$. A point $x^* \in X$ is called weakly Pareto optimal point if there is no $x \in X$ for which $F(x) < F(x^*)$. Similarly, a point $x^* \in X$ is a local Pareto optimal if there exists a neighborhood $Y \subseteq X$ of x^* such that the point x^* is a Pareto optima for F restricted on Y. Similarly, a point x^* is a local weak Pareto optima if there exists a neighborhood $Y \subseteq X$ of x^* such that the point x^* is a weak Pareto optimal for F restricted on Y. The matrix $JF(x) \in \mathbb{R}^{m \times n}$ is the Jacobian matrix of f_j at x, i.e., the j-th row of $JF(x)$ is $\nabla_q f_j(x)$ (q-gradient) for all $j = 1, \ldots, m$. Let $Wf_j(x)$ be the Hessian matrix of f_j at x for all $j = 1, \ldots, m$. Note that every Pareto optimal point is a weakly Pareto optimal point [37]. The directional derivative of f_j at x in the descent direction d_q is given as:

$$f'_j(x, d_q) = \lim_{\alpha \to 0} \frac{f_j(x + \alpha d_q) - f_j(x)}{\alpha} \quad (5)$$

The necessary condition to get the critical point for multiobjective optimization problems is given in [17]. For any $x \in \mathbb{R}^n$, $\|x\|$ denotes the Euclidean norm in $\mathbb{R}^n$. Let $K(x^0, r) = \{x : \|x - x^0\| \leq r\}$ with a center $x^0 \in \mathbb{R}^n$ and radius $r \in \mathbb{R}_+$. Norm of the matrix $A \in \mathbb{R}^{n \times n}$ is $\|A\| = \max_{x \in \mathbb{R}^{n \times n}} \frac{\|Ax\|}{\|x\|}, x \neq 0$. The following proposition indicates that when $f(x)$ is a linear function, then the q-gradient is similar to the classical gradient.

Proposition 1 ([33])**.** *If $f(x) = a + p^T x$, where $a \in \mathbb{R}$ and $p \in \mathbb{R}^n$, then for any $x \in \mathbb{R}^n$, and $q \in (0,1)$, we have $\nabla_q f(x) = \nabla f(x) = p$.*

All the quasi-Newton methods approximate the Hessian of function f as $W^k \in \mathbb{R}^{n \times n}$, and update the new formula based on previous approximation [38]. Line search methods are imperative methods for (UMOP) in which a search direction is first computed and then along this direction a step-length is chosen. The entire process is an iterative.

3. The q-Quasi-Newton Direction for Multiobjective

The most well-known quasi-Newton method for single objective function is the BFGS (Broyden, Fletcher, Goldfarb, and Shanno) method. This is a line search method along with a descent direction d_q^k within the context of q-derivative, given as:

$$d_q^k = -(W^k)^{-1} \nabla_q f(x^k), \tag{6}$$

where f is a continuously q-differentiable function, and $W^k \in \mathbb{R}^{n \times n}$ is a positive definite matrix that is updated at every iteration. The new point is:

$$x^{k+1} = x^k + \alpha_k d_q^k. \tag{7}$$

In the case of the Steepest Descent method and Newton's method, W^k is taken to be an Identity matrix and exact Hessian of f, respectively. The quasi-Newton BFGS scheme generates the next W^{k+1} as

$$W^{k+1} = W^k - \frac{W^k s^k (s^k)^T W^k}{(s^k)^T W^k s^k} + \frac{y^k (y^k)^T}{(s^k)^T y^k}, \tag{8}$$

where $s^k = x^{k+1} - x^k = \alpha_k d_q^k$, and $y^k = \nabla_q f(x^{k+1}) - \nabla_q f(x^k)$. In Newton's method, second-order differentiability of the function is required. While calculating W^k, we use q-derivative which behaves like a Hessian matrix of $f(x)$. W^{k+1} may not be a positive definite, which can be modified to be a positive definite through the symmetric indefinite factorization [39]. The q-Quasi-Newton's direction $d_q(x)$ is an optimal solution of the following modified problem [40] as:

$$\min_{d_q \in \mathbb{R}^n} \max_{j=1,\ldots,m} \nabla_q f_j(x) d_q + \frac{1}{2} d_q^T W_j(x)^T d_q, \tag{9}$$

where $W_j(x)$ is computed as (8). The solution and optimal value of (9) are:

$$\psi(x) = \min_{d_q \in \mathbb{R}^n} \max_{j=1,\ldots,m} \nabla_q f_j(x)^T d_q + \frac{1}{2} d_q^T W_j(x) d_q, \tag{10}$$

and

$$d_q(x) = \arg \min_{d_q \in \mathbb{R}^n} \max_{j=1,\ldots,m} \nabla f_j(x)^T d_q + \frac{1}{2} d_q^T W_j(x) d_q. \tag{11}$$

The problem (9) becomes a convex quadratic optimization problem (CQOP) as follows:

$$\begin{aligned}
\text{minimize} \quad & h(t, d_q) = t, \\
\text{subject to} \quad & \nabla_q f_j(x)^T d_q + \frac{1}{2} d_q^T W_j(x) d_q - t \leq 0, \; j = 1, \ldots, m, \\
\text{where} \quad & (t, d_q) \in \mathbb{R} \times \mathbb{R}^n.
\end{aligned} \tag{12}$$

The Lagrangian function of (CQOP) is:

$$L((t,d_q),\lambda) = t + \sum_{j=1}^{m} \lambda_j \left(\nabla_q f_j(x)^T d_q + \frac{1}{2} d_q^T W_j(x) d_q - t \right). \quad (13)$$

For $\lambda = (\lambda_1, \lambda_2, \ldots, \lambda_m)^T$, we obtain the following (KKT) conditions [40]:

$$\sum_{j=1}^{m} \lambda_j (\nabla_q f_j(x) + W_j(x) d_q) = 0, \quad (14)$$

$$\lambda_j \geq 0, \; j = 1, \ldots, m, \quad (15)$$

$$\sum_{j=1}^{m} \lambda_j = 1, \quad (16)$$

$$\nabla_q f_j(x)^T d_q + \frac{1}{2} d_q^T W_j(x) d_q \leq t, \; j = 1, \ldots, m, \quad (17)$$

$$\lambda_j \left(\nabla_q f_j(x)^T d_q + \frac{1}{2} d_q^T W_j(x) d_q - t \right) = 0, \; j = 1, \ldots, m. \quad (18)$$

The solution $(d_q(x), \psi(x))$ is unique, and set $\lambda_j = \lambda_j(x)$ for all $j = 1, \ldots, m$ with $d_q = d_q(x)$ and $t = \psi(x)$ for satisfying (14)–(18). From (14), we obtain

$$d_q(x) = -\left(\sum_{j=1}^{m} \lambda_j(x) W_j(x) \right)^{-1} \sum_{j=1}^{m} \lambda_j(x) \nabla_q f_j(x). \quad (19)$$

This is a so-called q-Quasi-Newton's direction for solving (UMOP). We present the basic result for relating the stationary condition at a given point x to its q-Quasi-Newton direction $d_q(x)$ and function ψ.

Proposition 1. *Let $\psi : X \to \mathbb{R}$ and $d_q : X \to \mathbb{R}^n$ be given by (10) and (11), respectively, and $W_j(x) \geq 0$ for all $x \in X$. Then,*

1. $\psi(x) \leq 0$ for all $x \in X$.
2. *The conditions below are equivalent:*

    (a) *The point x is non stationary.*
    (b) $d_q(x) \neq 0$
    (c) $\psi(x) < 0$.
    (d) $d_q(x)$ is a descent direction.

3. *The function ψ is continuous.*

Proof. Since $d_q = 0$, then from (10), we have

$$\psi(x) \leq \min_{d_q \in \mathbb{R}^n} \max_{j=1,\ldots,m} \nabla_q f_j(x)^T 0 + \frac{1}{2} d_q^T W_j(x) 0 = 0,$$

thus $\psi(x) \leq 0$. It means that $JF(x^*) d_q(x) \in \mathbb{R}_-^m$. Thus, the given point $x \in \mathbb{R}^n$ is non-stationary. Since $W_j(x)$ is positive definite, and from (10) and (11), we have

$$\nabla_q f_j(x)^T d_q(x) < \nabla f_j(x)^T d_q(x) + \frac{1}{2} d_q(x)^T W_j(x)^T d_q(x) = \psi(x) \leq 0.$$

Since $\psi(x)$ is the optimal value of (CQOP), and it is negative, thus solution of (CQOP) can never be $d_q(x) = 0$. It is sufficient to show that the continuity [41] of ψ in set $Y \subset X$. Since $\psi(x) \leq 0$, then

$$\nabla_q f_j(x)^T d_q(x) \leq -\frac{1}{2} d_q(x)^T W_j(x) d_q(x), \tag{20}$$

for all $j = 1, \ldots, m$, and $W_j(x)$, where $j = 1 \ldots, m$ are positive definite for all $x \in Y$. Thus, the eigenvalues of Hessian matrices $W_j(x)$, where $j = 1, \ldots, m$ are uniformly bounded away from zero on Y so there exists $R, S \in \mathbb{R}_+$ such that

$$R = \max_{x \in Y, j=1,\ldots,m} \|\nabla_q f_j(x)\|, \tag{21}$$

and

$$S = \min_{x \in Y, \|e\|=1, j=1,\ldots,m} e^T W_j(x) e. \tag{22}$$

From (20) and using Cauchy–Schwarz inequality, we get

$$\|\nabla_q f_j(x)\| \|d_q(x)\| \leq \frac{1}{2} S \|d_q(x)\|^2 \leq R \|d_q(x)\|,$$

that is,

$$d_q(x) \leq 2\frac{R}{S},$$

for all $x \in Y$, that is, Newton's direction is uniformly bounded on Y. We present the family of function $\{\aleph_{x,j}\}_{x \in Y, j=1,\ldots,m}$, where

$$\aleph_{x,j} : Y \to \mathbb{R},$$

and

$$z \to \nabla_q f(z)^T d_q(x) + \frac{1}{2} d_q(x)^T W_j(x) d_q(x).$$

We shall prove that this family of functions is uniformly equicontinuous. For small value $\epsilon_z \in \mathbb{R}_+$ there exists $\delta_z \in \mathbb{R}_+$, and for $y \in K(z, \delta_z)$, we have

$$\|W_j(y) - \nabla_q^2 f_j(z)\| < \frac{\epsilon_z}{2},$$

and

$$\|\nabla_q^2 f_j(y) - \nabla_q^2 f_j(z)\| < \frac{\epsilon_z}{2},$$

for all $j = 1, \ldots, m$. because of q-continuity of Hessian matrices, the second inequality is true. Since Y is compact space, then there exists a finite sub-cover.

$$\psi_{x,j}(z) = \nabla_q f_j(z)^T d_q(x) + \frac{1}{2} d_q(x)^T W_j(x) d_q(x),$$

that is

$$\psi_{x,j}(z) = \nabla_q f_j(z)^T d_q(x) + \frac{1}{2} d_q(x)^T \nabla^2 f_j(z) d_q(x) + \frac{1}{2} d_q(x)^T (W_j(z) - \nabla_q^2 f_j(z) d_q(x)).$$

To show the q-continuous of last term, set $y_1, y_2 \in Y$ such that $\|y_1 - y_2\| < \delta$ for small $\delta \in \mathbb{R}_+$, then

$$|\frac{1}{2}d_q(x)^T W_j(y_1) - \nabla_q^2 f_j(y_1) d_q(x) - \frac{1}{2}d_q(x)^T W_j(y_2) - \nabla_q^2 f_j(y_2) d_q(x)|$$

$$\leq \frac{1}{2}\|d_q(x)\|^2 (\|B_j(y_1) - \nabla^2 f_j(z_1))\| + \|\nabla_q^2 f_j(z_2)$$

$$- \nabla_q^2 f_j(z_1 + \|B_j(y_2) - \nabla^2 f_j(z_{21}))\| + \|\nabla_q^2 f_j(z_2) - \nabla_q^2 f_j(z_{21}\|)$$

$$\leq \frac{1}{2}\|d_q(x)\|^2 (\varepsilon_{z1} + \varepsilon_{z2}).$$

$\psi_{x,j}$ is uniformly continuous [40] for all $x \in Y$ and for all $j = 1, \ldots, m$. There exists $\delta \in \mathbb{R}_+$ such that for all $y, z \in Y$, $\|y - z\| < \delta$ implies $|\psi(y) - \psi(z)| < \epsilon$ for all $x \in Y$. Thus, $\|y - z\| < \delta$.

$$\psi(z) \leq \max_{j=1,\ldots,m} \nabla f_j(z)^T d_q(y) + \frac{1}{2}d_q(y)^T W_j(z) d_q(y) = \phi_y(z)$$

$$\leq \phi_y(y) + |\phi_y(z) - \phi_y(y)| < \psi(y) + \epsilon.$$

Thus, $\psi(z) - \psi(y) < \epsilon$. If we interchange y and z, then $|\psi(z) - \psi(y)| < \epsilon$. It proves the continuity of ψ. □

The following modified lemma is due to [17,42].

Lemma 1. *Let $F : \mathbb{R}^n \to \mathbb{R}^m$ be continuously q-differentiable. If $x^* \in X$ is not a critical point for $\nabla_q(x) d_q < 0$, where $d_q \in \mathbb{R}^n$, $\sigma \in (0, 1]$, and $\varepsilon > 0$. Then,*

$$x + \alpha d_q(x) \in X \text{ and } F(x + \alpha d_q(x)) < F(x) + \alpha \gamma \psi(x),$$

for any $\alpha \in (0, \sigma]$ and $\gamma \in (0, \varepsilon]$.

Proof. Since x^* is not a critical point, then $\psi(x) < 0$. Let $r > 0$ such that $B(x, r) \subset X$ and $\alpha \in (0, \sigma]$. Therefore,

$$F(x + \alpha d_q(x)) - F(x) = \alpha \nabla_q F(x)^T d_q(x) + o_j(\alpha d_q(x), x)$$

Since $\nabla_q(x) d_q(x) < \psi(x)$, for $\alpha \in (0, \sigma]$, then

$$F(x + \alpha d_q(x)) - F(x) = \alpha \gamma \psi(x) + \alpha(1 - \sigma) \psi(x) + o_j(\alpha d_q(x), x).$$

The last term in the right-hand side of the above equation is non-positive because $\psi(x) \leq \frac{\psi(x^*)}{2} < 0$, for $\alpha \in [0, \sigma]$. □

4. Algorithm and Convergence Analysis

We first present the following Algorithm 1 [43] to find the gradient of the function using q-calculus. The higher-order q-derivative of f can be found in [44].

Algorithm 1 q-Gradient Algorithm

1: Input $q \in (0, 1), f(x), x \in \mathbb{R}, z$.
2: **if** $x = 0$ **then**
3: Set $g \leftarrow \lim \left(\frac{f(z) - f(q*z)}{(z - q*z)}, z, 0 \right)$.
4: **else**
5: Set $g \leftarrow \frac{f(x) - f(q*x)}{(x - q*x)}$.
6: Print $\nabla_q f(x) \leftarrow g$.

Example 1. *Given that $f : \mathbb{R}^2 \to \mathbb{R}$ defined by $f(x_1, x_2) = x_2^2 + 3x_1^3$. Then $\nabla_q f(x) = \begin{bmatrix} 3x_1^2(1+q+q^2) \\ x_2(1+q) \end{bmatrix}$.*

We are now prepared to write the unconstrained q-Quasi-Newton's Algorithm 2 for solving (UMOP). At each step, we solve the (CQOP) to find the q-Quasi-Newton direction. Then, we obtain the step length using the Armijo line search method. In every iteration, the new point and Hessian approximation are generated based on historical values.

Algorithm 2 q-Quasi-Newton's Algorithm for Unconstrained Multiobjective (q-QNUM)

1: Choose $q \in (0,1), x^0 \in X$, symmetric definite matrix $W^0 \in \mathbb{R}^{n \times n}, c \in (0,1)$, and a small tolerance value $\epsilon > 0$.
2: **for** k=0,1,2,... **do**
3:     Solve (CQOP).
4:     Compute d_q^k and ψ^k.
5:     **if** $\psi^k > -\epsilon$ **then**
6:         Stop.
7:     **else**
8:         Choose α_k as the $\alpha \in (0,1]$ such that $x^k + \alpha d_q^k \in X$ and $F(x^k + \alpha d_q^k) \le F(x^k) + c\alpha\psi^k$.
9:         Update $x^{k+1} \leftarrow x^k + \alpha_k d_q^k$.
10:        Update W_j^{k+1}, where $j = 1, \ldots, m$ using (8).

We now finally start to show that every sequence produced by the proposed method converges to a weakly efficient point. It does not matter how poorly the initial point is guessed. We assume that the method does not stop, and produces an infinite sequence of iterates. We now present the modified sufficient conditions for the superlinear convergence [17,40] within the context of q-calculus.

Theorem 1. *Let $\{x^k\}$ be a sequence generated by (q-QNUM), and $Y \subset X$ be a convex set. Also, $\gamma \in (0,1)$ and $r, a, b, \delta, \epsilon > 0$, and*

(a) $aI \le W_j(x) \le bI$ for all $x \in Y, j = 1, \ldots, m$,
(b) $\|\nabla_q^2 f_j(y) - \nabla_q^2 f_j(x),\| < \frac{\epsilon}{2}$ for all $x,y \in Y$ with $\|y-x\| \in \delta$,
(c) $\|(W_j^k - \nabla_q^2 f_j(x^k))(y - x^k)\| < \frac{\epsilon}{2}\|y - x^k\|$ for all $k \ge k_0, y \in Y, j = 1, \ldots, m$,
(d) $\frac{\epsilon}{a} \le 1 - c$,
(e) $B(x^0, r) \in Y$,
(f) $\|d_q(x^0)\| < \min\{\delta, r(1 - \frac{\epsilon}{a})\}$.

Then, for all $k \ge k_0$, we have that

1. $\|x^k - x^{k_0}\| \le \|d_q(x^0)\| \frac{1 - (\frac{\epsilon}{a})^{k-k_0}}{1 - (\frac{\epsilon}{a})}$
2. $\alpha_k = 1$,
3. $\|d_q(x^k)\| \le \|d_q(x^{k_0})\|(\frac{\epsilon}{a})^{k-k_0}$,
4. $\|d_q(x^{k+1})\| \le \|d_q(x^k)\|\frac{\epsilon}{a}$.

Then, the sequence $\{x^k\}$ converges to local Pareto points $x^* \in \mathbb{R}^m$, and the convergence rate is superlinear.

Proof. From part 1, part 3 of this theorem and triangle inequality,

$$\|x^k + d_q(x^k) - x^0\| \le \frac{1 - (\frac{\epsilon}{a})^{k+1}}{1 - \frac{\epsilon}{a}} \|d_q(x^{k_0})\|.$$

From (d) and (f), we follow $x^k, x^k + d_q(x^k) \in K(x^{k_0}, r)$ and $x^k + d_q(x^k) - x^k < \delta$. We also have

$$f_j(x^k + d_q(x^k)) \le f_j(x^k) + d_q(x^k)\nabla_q f(x^k) + \frac{1}{2}d_q(x^k)(\nabla_q^2 f)(x^k) + \frac{1}{2}\|d_q(x^k)\|^2,$$

that is,
$$f_j(x^k + d_q(x^k)) \le f_j(x^k) + \psi(x^k) + \frac{\varepsilon}{2}\|d_q(x^k)\|^2$$
$$= f_j(x^k) + \gamma\psi(x^k) + (1-\gamma)\psi(x^k) + \frac{\varepsilon}{2}\|d_q(x^k)\|^2.$$

Since $\psi \le 0$ and $(1-\gamma)\psi(x^k) + \frac{\varepsilon}{2}\|d_q(x^k)\|^2 \le (\varepsilon - a(1-\gamma))\frac{\|d_q(x^k)\|^2}{2} \le 0$, we get
$$f_j(x^k + d_q(x^k)) \le f_j(x^k) + \gamma\psi(x^k),$$

for all $j = 1,\ldots,m$. The Armijo conditions holds for $\alpha_k = 1$. Part 1 of this theorem holds. We now set $x^k, x^{k+1} \in K(x^{k_0}, r)$, and $\|x^{k+1} - x^k\| < \delta$. Thus, we get $x^{k+1} = x^k + d_q(x^k)$. We now define $v(x^{k+1}) = \sum_{j=1}^m \lambda_j^k \nabla_q f_j(x^{k+1})$. Therefore,
$$|\psi(x^{k+1})| \le \frac{1}{2a}\|v(x^{k+1})\|^2.$$

We now estimate $\|v(x^{k+1})\|$. For $x \in X$, we define
$$G^k(x) := \sum_{j=1}^m \lambda_j^k f_j(x^{k+1}),$$

and
$$H^k = \sum_{j=1}^m \lambda_j^k W_j(x^k),$$

where $\lambda_j^k \ge 0$, for all $j = 1,\ldots,m$, are KKT multipliers. We obtain following:
$$\nabla_q G^k(x) = \sum_{j=1}^m \lambda_j^k \nabla_q f_j(x),$$

and
$$\nabla_q^2 G^k(x) = \sum_{j=1}^m \lambda_j^k \nabla_q^2 f_j(x).$$

Then, $v^{k+1} = \nabla_q G^k(x^{k+1})$. We get
$$d_q(x^k) = -(H^k)^{-1}\nabla_q G^k(x^k).$$

From assumptions (b) and (c) of this theorem,
$$\|\nabla_q^2 G^k(y) - \nabla_q^2 G^k(x^k)\| < \frac{\varepsilon}{2},$$
$$\|(H^k - \nabla_q^2 G^k(x^k))(y - x^k)\| < \frac{\varepsilon}{2}\|y - x^k\|$$
hold for all $x, y \in Y$ with $\|y - x\| < \delta$ and $k \ge k_0$. We have
$$\nabla_q G^k(x^k + d_q(x^k)) - (\nabla_q G^k(x^k) + H^k d_q(x^k))\| < \epsilon\|d_q(x^k)\|.$$

Since $\nabla_q G^k(x^k) + H^k d_q(x^k) = 0$, then
$$\|v(x^{k+1})\| = \|\nabla_q G^k(x^{k+1})\| < \epsilon\|d_q(x^k)\|,$$

and
$$|\psi^{k+1}| \leq \frac{1}{2a}\|v(x^{k+1})\|^2 < \frac{\epsilon^2}{2a}\|d_q(x^k)\|^2.$$

We have
$$\frac{a}{2}\|d_q(x^{k+1})\|^2 < \frac{\epsilon^2}{2a}\|d_q(x^k)\|^2.$$

Thus,
$$\|d_q(x^{k+1})\| < \frac{\epsilon}{a}\|d_q(x^k)\|$$

Thus, part 4 is proved. We finally prove superlinear convergence of $\{x^k\}$. First we define
$$r^k = \|d_q^0\|\frac{\frac{\bar{\epsilon}^{k-k_0}}{a}}{1-\frac{\bar{\epsilon}}{a}},$$

and
$$\delta^k = \|d_q^{k_0}\|\left(\frac{\bar{\epsilon}}{a}\right)^{k-k_0}.$$

From triangle inequality, assumptions (e), (f) and part 1, we have $K(x^k, r^k) \subset K(x^{k_0}, r) \subset V$. Choose any $\tau \in \mathbb{R}_+$, and define
$$\bar{\epsilon} = \min\{a\frac{\tau}{1+2\tau}, \epsilon\}.$$

For $k \geq k_0$ inequalities
$$\|\nabla_q^2 f_j(y) - \nabla_q^2 f_j(x)\| < \frac{\bar{\epsilon}}{2}$$

for all $x, y \in K(x^k, r^k)$ with $\|y - x\| < \delta^k$, and
$$\|W_j(x^l) - \nabla_q^2 f_j(x^l)(y - x^l)\| < \frac{\bar{\epsilon}}{2}$$

for all $y \in K(x^k, r^k)$ and $l \geq k$ holds both for $j = 1, \ldots, m$. Assumptions (a)–(f) are satisfied for $\bar{\epsilon}, r^k, \delta^k$, and x^k instead of ϵ, r, δ, and x^0, respectively. We have
$$\|x^l - x^k\| \leq \|d_q(x^k)\|\frac{1-\left(\frac{\bar{\epsilon}}{a}\right)^{l-k}}{1-\frac{\bar{\epsilon}}{a}}.$$

Let $l \to \infty$ and we get $\|x^* - x^k\| \leq \|d_q(x^k)\|\frac{1}{1-\frac{\bar{\epsilon}}{a}}$. Using the last inequality, and part 4, we have
$$\|x^* - x^{k+1}\| \leq \|d_q(x^{k+1})\|\frac{1}{1-\frac{\bar{\epsilon}}{a}} \leq \|d_q(x^k)\|\frac{\frac{\bar{\epsilon}}{a}}{1-\frac{\bar{\epsilon}}{a}}.$$

From above and triangle inequality, we have
$$\|x^* - x^{k+1}\| \geq \|x^{k+1} - x^k\| - \|x^* - x^{k+1}\|,$$

that is,
$$\|x^* - x^{k+1}\| \geq \|d_q(x^k)\| - \|d^k\|\frac{\bar{\epsilon}}{1-\frac{\bar{\epsilon}}{a}} = \|d_q^k\|\frac{1-2\frac{\bar{\epsilon}}{a}}{1-\frac{\bar{\epsilon}}{a}}. \qquad (23)$$

Since $1 - 2\frac{\bar{\epsilon}}{a} > 0$, and $1 - 2\frac{\bar{\epsilon}}{a} > 0$, then we get
$$\|x^* - x^{k+1}\| \leq \tau\|x^* - x^k\|,$$

where $\tau \in \mathbb{R}_+$ is chosen arbitrarily. Thus, the sequence $\{x^k\}$ converges superlinearly to x^*. □

5. Numerical Results

The proposed algorithm (q-QNUM), i.e., Algorithm 2, presented in Section 4 is implemented in MATLAB (2017a) and tested on some test problems known from the literature. All tests were run under the same conditions. The box constraints of the form $lb \leq x \leq ub$ are used for each test problem. These constraints are considered under the direction search problem (CQOP) such that the newly generated point always lies in the same box, that is, $lb \leq x + d_q \leq ub$ holds. We use the stopping criteria at x^k as: $\psi(x^k) > -\epsilon$ where $\epsilon \in \mathbb{R}_+$. All test problems given in Table 1 are solved 100 times. The starting points are randomly chosen from a uniform distribution between lb and ub. The first column in the given table is the name of the test problem. We use the abbreviation of author's names and number of the problem in the corresponding paper. The second column indicates the source of the paper. The third column is for lower bound and upper bound. We compare the results of (q-QNUM) with (QNMO) of [40] in the form of a number of iterations (*iter*), number of objective functions evaluation (*obj*), and number of gradient evaluations (*grad*), respectively. From Table 1, we can conclude that our algorithm shows better performance.

Example 2. *Find the approximate Pareto front using (q-QNUM) and (QNMO) for the given (UMOP) [45]:*

$$\text{Minimize } f_1(x_1, x_2) = (x_1 - 1)^2 + (x_1 - x_2)^2,$$
$$\text{Minimize } f_2(x_1, x_2) = (x_2 - 3)^2 + (x_1 - x_2)^2,$$

where $-3 \leq x_1, x_2 \leq 10$.

The number of Pareto points generated due to (q-QNUM) with Algorithm 1 and (QNMO) is shown in Figure 1. One can observe that the number of iterations as $iter = 200$ in (q-QNUM) and $iter = 525$ in (QNMO) are responsible for generating the approximate Pareto front of above (UMOP).

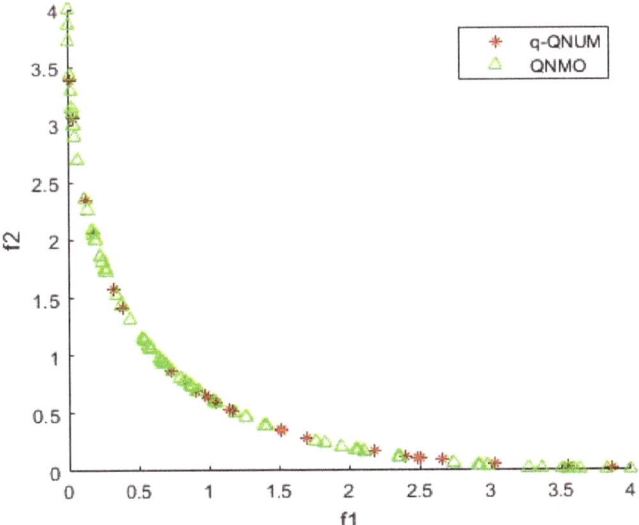

Figure 1. Approximate Pareto Front of Example 1.

Table 1. Numerical Results of Test Problems.

Problem	Source	[lb,ub]	(q-QNUM)			(QNMO)		
			iter	obj	grad	iter	obj	grad
BK1	[46]	[−5, 10]	200	200	200	200	200	200
MOP5	[46]	[−30, 30]	141	965	612	333	518	479
MOP6	[46]	[0, 1]	250	2177	1712	181	2008	2001
MOP7	[46]	[−400, 400]	200	200	200	751	1061	1060
DG01	[47]	[−10, 13]	175	724	724	164	890	890
IKK1	[47]	[−50, 50]	170	170	170	253	254	253
SP1	[45]	[−3, 2]	200	200	200	525	706	706
SSFYY1	[45]	[−2, 2]	200	200	200	200	300	300
SSFYY2	[45]	[−100, 100]	263	277	277	263	413	413
SK1	[48]	[−10, 10]	139	1152	1152	87	732	791
SK2	[48]	[−3, 11]	154	1741	1320	804	1989	1829
VU1	[49]	[−3, 3]	316	1108	1108	11,361	19,521	11,777
VU2	[49]	[−3, 7]	99	1882	1882	100	1900	1900
VFM1	[50]	[−2, 2]	195	195	195	195	290	290
VFM2	[50]	[−4, 4]	200	200	200	524	693	678
VFM3	[50]	[−3, 3]	161	1130	601	690	1002	981

6. Conclusions

The q-Quasi-Newton method converges superlinearly to the solution of (UMOP) if all objective functions are strongly convex within the context of q-derivative. In a neighborhood of this solution, the algorithm uses a full Armijo steplength. The numerical performance of the proposed algorithm is faster than their actual evaluation.

Author Contributions: K.K.L. gave reasonable suggestions for this manuscript; S.K.M. gave the research direction of this paper; B.R. revised and completed this manuscript. All authors have read and agreed to the published version of the manuscript.

Funding: This research was supported by the Science and Engineering Research Board (Grant No. DST-SERB-MTR-2018/000121) and the University Grants Commission (IN) (Grant No. UGC-2015-UTT–59235).

Acknowledgments: The authors are grateful to the anonymous reviewers and the editor for the valuable comments and suggestions to improve the presentation of this paper.

Conflicts of Interest: The authors declare no conflict of interest.

References

1. Eschenauer, H.; Koski, J.; Osyczka, A. *Multicriteria Design Optimization: Procedures and Applications*; Springer: Berlin, Germany, 1990.
2. Haimes, Y.Y.; Hall, W.A.; Friedmann, H.T. *Multiobjective Optimization in Water Resource Systems, The Surrogate Worth Trade-off Method*; Elsevier Scientific: Amsterdam, The Netherlands, 1975.
3. Nwulu, N.I.; Xia, X. Multi-objective dynamic economic emission dispatch of electric power generation integrated with game theory based demand response programs. *Energy Convers. Manag.* **2015**, *89*, 963–974. [CrossRef]
4. Badri, M.A.; Davis, D.L.; Davis, D.F.; Hollingsworth, J. A multi-objective course scheduling model: Combining faculty preferences for courses and times. *Comput. Oper. Res.* **1998**, *25*, 303–316. [CrossRef]

5. Ishibuchi, H.; Nakashima, Y.; Nojima, Y. Performance evaluation of evolutionary multiobjective optimization algorithms for multiobjective fuzzy genetics-based machine learning. *Soft Comput.* **2011**, *15*, 2415–2434. [CrossRef]
6. Liu, S.; Vicente, L.N. The stochastic multi-gradient algorithm for multi-objective optimization and its application to supervised machine learning. *arXiv* **2019**, arXiv:1907.04472.
7. Tavana, M. A subjective assessment of alternative mission architectures for the human exploration of mars at NASA using multicriteria decision making. *Comput. Oper. Res.* **2004**, *31*, 1147–1164. [CrossRef]
8. Gass, S.; Saaty, T. The computational algorithm for the parametric objective function. *Nav. Res. Logist. Q.* **1955**, *2*, 39–45. [CrossRef]
9. Miettinen, K. *Nonlinear Multiobjective Optimization*; Kluwer Academic: Boston, MA, USA, 1999.
10. Fishbum, P.C. Lexicographic orders, utilities and decision rules: A survey. *Manag. Sci.* **1974**, *20*, 1442–1471. [CrossRef]
11. Coello, C.A. An updated survey of GA-based multiobjective optimization techniques. *ACM Comput. Surv. (CSUR)* **2000**, *32*, 109–143. [CrossRef]
12. Fliege, J.; Svaiter, B.F. Steepest descent method for multicriteria optimization. *Math. Method. Oper. Res.* **2000**, *51*, 479–494. [CrossRef]
13. Drummond, L.M.G.; Iusem, A.N. A projected gradient method for vector optimization problems. *Comput. Optim. Appl.* **2004**, *28*, 5–29. [CrossRef]
14. Drummond, L.M.G.; Svaiter, B.F. A steepest descent method for vector optimization. *J. Comput. Appl. Math.* **2005**, *175*, 395–414. [CrossRef]
15. Branke, J.; Dev, K.; Miettinen, K.; Slowiński, R. (Eds.) *Multiobjective Optimization: Interactive and Evolutionary Approaches*; Springer: Berlin, Germany, 2008.
16. Mishra, S.K.; Ram, B. *Introduction to Unconstrained Optimization with R*; Springer Nature: Singapore, 2019; pp. 175–209.
17. Fliege, J.; Drummond, L.M.G.; Svaiter, B.F. Newton's method for multiobjective optimization. *SIAM J. Optim.* **2009**, *20*, 602–626. [CrossRef]
18. Chuong, T.D. Newton-like methods for efficient solutions in vector optimization. *Comput. Optim. Appl.* **2013**, *54*, 495–516. [CrossRef]
19. Qu, S.; Liu, C.; Goh, M.; Li, Y.; Ji, Y. Nonsmooth Multiobjective Programming with Quasi-Newton Methods. *Eur. J. Oper. Res.* **2014**, *235*, 503–510. [CrossRef]
20. Jiménez; M.A.; Garzón, G.R.; Lizana, A.R. (Eds.) *Optimality Conditions in Vector Optimization*; Bentham Science Publishers: Sharjah, UAE, 2010.
21. Al-Saggaf, U.M.; Moinuddin, M.; Arif, M.; Zerguine, A. The q-least mean squares algorithm. *Signal Process.* **2015**, *111*, 50–60. [CrossRef]
22. Aral, A.; Gupta, V.; Agarwal, R.P. *Applications of q-Calculus in Operator Theory*; Springer: New York, NY, USA, 2013.
23. Rajković, P.M.; Marinković, S.D.; Stanković, M.S. Fractional integrals and derivatives in q-calculus. *Appl. Anal. Discret. Math.* **2007**, *1*, 311–323.
24. Gauchman, H. Integral inequalities in q-calculus. *Comput. Math. Appl.* **2004**, *47*, 281–300. [CrossRef]
25. Bangerezako, G. Variational q-calculus. *J. Math. Anal. Appl.* **2004**, *289*, 650–665. [CrossRef]
26. Abreu, L. A q-sampling theorem related to the q-Hankel transform. *Proc. Am. Math. Soc.* **2005**, *133*, 1197–1203. [CrossRef]
27. Koornwinder, T.H.; Swarttouw, R.F. On q-analogues of the Fourier and Hankel transforms. *Trans. Am. Math. Soc.* **1992**, *333*, 445–461.
28. Ernst, T. *A Comprehensive Treatment of q-Calculus*; Springer: Basel, Switzerland; Heidelberg, Germany; New York, NY, USA; Dordrecht, The Netherlands; London, UK, 2012.
29. Noor, M.A.; Awan, M.U.; Noor, K.I. Some quantum estimates for Hermite-Hadamard inequalities. *Appl. Math. Comput.* **2015**, *251*, 675–679. [CrossRef]
30. Pearce, C.E.M.; Pečarić, J. Inequalities for differentiable mappings with application to special means and quadrature formulae. *Appl. Math. Lett.* **2000**, *13*, 51–55. [CrossRef]
31. Ernst, T. A Method for q-Calculus. *J. Nonl. Math. Phys.* **2003**, *10*, 487–525. [CrossRef]

32. Sterroni, A.C.; Galski, R.L.; Ramos, F.M. The q-gradient vector for unconstrained continuous optimization problems. In *Operations Research Proceedings*; Hu, B., Morasch, K., Pickl, S., Siegle, M., Eds.; Springer: Heidelberg, Germany, 2010; pp. 365–370.
33. Gouvêa, E.J.C.; Regis, R.G.; Soterroni, A.C.; Scarabello, M.C.; Ramos, F.M. Global optimization using q-gradients. *Eur. J. Oper. Res.* **2016**, *251*, 727–738. [CrossRef]
34. Chakraborty, S.K.; Panda, G. Newton like line search method using q-calculus. In *International Conference on Mathematics and Computing. Communications in Computer and Information Science*; Giri, D., Mohapatra, R.N., Begehr, H., Obaidat, M., Eds.; Springer: Singapore, 2017; Volume 655, pp. 196–208.
35. Mishra, S.K.; Panda, G.; Ansary, M.A.T.; Ram, B. On q-Newton's method for unconstrained multiobjective optimization problems. *J. Appl. Math. Comput.* **2020**. [CrossRef]
36. Jackson, F.H. On q-functions and a certain difference operator. *Earth Environ. Sci. Trans. R. Soc. Edinb.* **1908**, *46*, 253–281. [CrossRef]
37. Bento, G.C.; Neto, J.C. A subgradient method for multiobjective optimization on Riemannian manifolds. *J. Optimiz. Theory App.* **2013**, *159*, 125–137. [CrossRef]
38. Andrei, N. A diagonal quasi-Newton updating method for unconstrained optimization. *Numer. Algorithms* **2019**, *81*, 575–590. [CrossRef]
39. Nocedal, J.; Wright, S.J. *Numerical Optimization*, 2nd ed.; Springer Series in Operations Research and Financial Engineering; Springer: New York, NY, USA, 2006.
40. Povalej, Z. Quasi-Newton's method for multiobjective optimization. *J. Comput. Appl. Math.* **2014**, *255*, 765–777. [CrossRef]
41. Ye, Y.L. D-invexity and optimality conditions. *J. Math. Anal. Appl.* **1991**, *162*, 242–249. [CrossRef]
42. Morovati, V.; Basirzadeh, H.; Pourkarimi, L. Quasi-Newton methods for multiobjective optimization problems. *4OR-Q. J. Oper. Res.* **2018**, *16*, 261–294. [CrossRef]
43. Samei, M.E.; Ranjbar, G.K.; Hedayati, V. Existence of solutions for equations and inclusions of multiterm fractional q-integro-differential with nonseparated and initial boundary conditions. *J. Inequal Appl.* **2019**, *273*. [CrossRef]
44. Adams, C.R. The general theory of a class of linear partial difference equations. *Trans. Am. Math.Soc.* **1924**, *26*, 183–312.
45. Sefrioui, M.; Perlaux, J. Nash genetic algorithms: Examples and applications. In Proceedings of the 2000 Congress on Evolutionary Computation, La Jolla, CA, USA, 16–19 July 2000; Volume 1, pp. 509–516.
46. Huband, S.; Hingston, P.; Barone, L.; While, L. A review of multiobjective test problems and a scalable test problem toolkit. *IEEE T. Evolut. Comput.* **2006**, *10*, 477–506. [CrossRef]
47. Ikeda, K.; Kita, H.; Kobayashi, S. Failure of Pareto-based MOEAs: Does non-dominated really mean near to optimal? In Proceedings of the 2001 Congress on Evolutionary Computation, Seoul, Korea, 27–30 May 2001; Volume 2, pp. 957–962.
48. Shim, M.B.; Suh, M.W.; Furukawa, T.; Yagawa, G.; Yoshimura, S. Pareto-based continuous evolutionary algorithms for multiobjective optimization. *Eng Comput.* **2002**, *19*, 22–48. [CrossRef]
49. Valenzuela-Rendón, M.; Uresti-Charre, E.; Monterrey, I. A non-generational genetic algorithm for multiobjective optimization. In Proceedings of the Seventh International Conference on Genetic Algorithms, East Lansing, MI, USA, 19–23 July 1997; pp. 658–665.
50. Vlennet, R.; Fonteix, C.; Marc, I. Multicriteria optimization using a genetic algorithm for determining a Pareto set. *Int. J. Syst. Sci.* **1996**, *27*, 255–260. [CrossRef]

© 2020 by the authors. Licensee MDPI, Basel, Switzerland. This article is an open access article distributed under the terms and conditions of the Creative Commons Attribution (CC BY) license (http://creativecommons.org/licenses/by/4.0/).

Article

Convergence Analysis and Complex Geometry of an Efficient Derivative-Free Iterative Method

Deepak Kumar [1,2,*,†], **Janak Raj Sharma** [1,*,†] **and Lorentz Jäntschi** [3,4,*]

1. Department of Mathematics, Sant Longowal Institute of Engineering and Technology, Longowal 148106, Sangrur, India
2. Chandigarh University, Gharuan 140413, Mohali, India
3. Department of Physics and Chemistry, Technical University of Cluj-Napoca, Cluj-Napoca 400114, Romania
4. Institute of Doctoral Studies, Babeș-Bolyai University, Cluj-Napoca 400084, Romania
* Correspondence: deepak.babbi@gmail.com (D.K.); jrshira@yahoo.co.in (J.R.S.); lorentz.jantschi@gmail.com (L.J.)
† These authors contributed equally to this work.

Received: 12 September 2019; Accepted: 29 September 2019; Published: 2 October 2019

Abstract: To locate a locally-unique solution of a nonlinear equation, the local convergence analysis of a derivative-free fifth order method is studied in Banach space. This approach provides radius of convergence and error bounds under the hypotheses based on the first Fréchet-derivative only. Such estimates are not introduced in the earlier procedures employing Taylor's expansion of higher derivatives that may not exist or may be expensive to compute. The convergence domain of the method is also shown by a visual approach, namely basins of attraction. Theoretical results are endorsed via numerical experiments that show the cases where earlier results cannot be applicable.

Keywords: local convergence; nonlinear equations; Banach space; Fréchet-derivative

MSC: 49M15; 47H17; 65H10

---

1. Introduction

Banach [1] or complete normed vector spaces constantly bring new solving strategies for real problems in domains dealing with numerical methods (see for example [2–5]). In this context, development of new methods [6] and their convergence analysis [7] are of growing interest.

Let B_1, B_2 be Banach spaces and $\Omega \subseteq B_1$ be closed and convex. In this study, we locate a solution x^* of the nonlinear equation

$$F(x) = 0, \qquad (1)$$

where $F : \Omega \subseteq B_1 \to B_2$ is a Fréchet-differentiable operator. In computational sciences, many problems can be transformed into form (1). For example, see the References [8–11]. The solution of such nonlinear equations is hardly attainable in closed form. Therefore, most of the methods for solving such equations are usually iterative. The important issue addressed to an iterative method is its domain of convergence since it gives us the degree of difficulty for obtaining initial points. This domain is generally small. Thus, it is necessary to enlarge the domain of convergence but without any additional hypotheses. Another important problem related to convergence analysis of an iterative method is to find precise error estimates on $\|x_{n+1} - x_n\|$ or $\|x_n - x^*\|$.

A good reference for the general principles of functional analysis is [12]. Recurrence relations for rational cubic methods are revised in [13] (for Halley method) and in [14] (for Chebyshev method). A new iterative modification of Newton's method for solving nonlinear scalar equations was proposed in [15], while a modification of a variant of it with accelerated third order convergence was proposed in [16]. An ample collection of iterative methods is found in [9]. The recurrence relations for Chebyshev-type methods accelerating classical Newton iteration have been introduced in [17], recurrence relations in a third-order family of Newton-like methods for approximating solution of a nonlinear equation in Banach spaces were studied in [18]. In the context of Kantrovich assumptions for semilocal convergence of a Chebyshev method, the convergence conditions are significantly reduced in [19]. The computational efficiency and the domain of the uniqueness of the solution were readdressed in [20]. The point of attraction of two fourth-order iterative Newton-type methods was studied in [21], while convergence ball and error analysis of Newton-type methods with cubic convergence were studied in [22,23]. Weaker conditions for the convergence of Newton's method are given in [24], while further analytical improvements in two particular cases as well as numerical analysis in the general case are given in [25], while local convergence of three-step Newton–Gauss methods in Banach spaces was recently analyzed in [26]. Recently, researchers have also constructed some higher order methods; see, for example [27–31] and references cited therein.

One of the basic methods for approximating a simple solution x^* of Equation (1) is the quadratically convergent derivative-free Traub–Steffensen's method, which is given by

$$x_{n+1} = M_{2,1}(x_n) = x_n - [u_n, x_n; F]^{-1} F(x_n), \text{ for each } n = 0, 1, 2, \ldots, \tag{2}$$

where $u_n = x_n + \beta F(x_n)$, $\beta \in \mathbb{R} - \{0\}$ has a quadratic order of convergence. Based on (2), Sharma et al. [32] have recently proposed a derivative-free method with fifth order convergence for approximating a solution of $F(x) = 0$ using the weight-function scheme defined for each $n = 0, 1 \ldots$ by

$$\begin{aligned} y_n &= M_{2,1}(x_n), \\ z_n &= y_n - [u_n, x_n; F]^{-1} F(y_n), \\ x_{n+1} &= z_n - H(x_n)[u_n, x_n; F]^{-1} F(z_n), \end{aligned} \tag{3}$$

wherein $H(x_n) = 2I - [u_n, x_n; F]^{-1}[z_n, y_n; F]$. The computational efficiency of this method was discussed in detail and performance was favorably compared with existing methods in [32]. To prove the local convergence order, the authors used Taylor expansions with hypotheses based on a Fréchet-derivative up to the fifth order. It is quite clear that these hypotheses restrict the applicability of methods to the problems involving functions that are at least five times Fréchet-differentiable. For example, let us define a function g on $\Omega = [-\frac{1}{2}, \frac{5}{2}]$ by

$$g(t) = \begin{cases} t^3 \ln t^2 + t^5 - t^4, & t \neq 0, \\ 0, & t = 0. \end{cases} \tag{4}$$

We have that

$$g'(t) = 3t^2 \ln t^2 + 5t^4 - 4t^3 + 2t^2,$$
$$g''(t) = 6t \ln t^2 + 20t^3 - 12t^2 + 10t$$

and

$$g'''(t) = 6 \ln t^2 + 60t^2 - 24t + 22.$$

Then, g''' is unbounded on Ω. Notice also that the proofs of convergence use Taylor expansions.

In this work, we study the local convergence of the methods (3) using the hypotheses on the first Fréchet-derivative only taking advantage of the Lipschitz continuity of the first Fréchet-derivative. Moreover, our results are presented in the more general setting of a Banach space. We summarize the contents of the paper. In Section 2, the local convergence analysis of method (3) is presented. In Section 3, numerical examples are performed to verify the theoretical results. Basins of attraction showing convergence domain are drawn in Section 4. Concluding remarks are reported in Section 5.

2. Local Convergence Analysis

Let's study the local convergence of method (3). Let $p \geq 0$ and $M \geq 0$ be the parameters and $w_0 : [0, +\infty)^2 \to [0, +\infty)$ be a continuous and nondecreasing function with $w_0(0,0) = 0$. Let the parameter r be defined by

$$r = \sup\{t \geq 0;\, w_0(pt, t) < 1\}. \tag{5}$$

Consider the functions $w_1 : [0, r)^2 \to [0, +\infty)$ and $v_0 : [0, r) \to [0, +\infty)$ as continuous and nondecreasing. Furthermore, define functions g_1 and h_1 on the interval $[0, r)$ as

$$g_1(t) = \frac{w_1(\beta v_0(t)t, t)}{1 - w_0(pt, t)}$$

and

$$h_1(t) = g_1(t) - 1.$$

Suppose that

$$w_1(0, 0) < 1. \tag{6}$$

From (6), we obtain that

$$h_1(0) = \frac{w_1(0,0)}{1 - w_0(0,0)} - 1 < 0$$

and, by (5), $h_1(t) \to +\infty$ as $t \to r^-$. Then, it follows from the intermediate value theorem [33] that equation $h_1(t) = 0$ has solutions in $(0, r)$. Denote by r_1 the smallest such solution.
Furthermore, define functions g_2 and h_2 on the interval $[0, r_1)$ by

$$g_2(t) = \left(1 + \frac{M}{1 - w_0(pt, t)}\right) g_1(t)$$

and

$$h_2(t) = g_2(t) - 1.$$

Then, we have that $h_2(0) = -1 < 0$ and $h_2(t) \to +\infty$ as $t \to r_1^-$. Let r_2 be the smallest zero of function h_2 on the interval $(0, r_1)$.
Finally, define the functions $\tilde{g}$, g_3 and h_3 on the interval $[0, r_2)$ by

$$\tilde{g}(t) = \frac{1}{1 - w_0(pt, t)}\left(1 + \frac{(w_0(pt, t) + w_0(g_1(t)t, g_2 t))}{1 - w_0(pt, t)}\right),$$

$$g_3(t) = (1 + M\tilde{g}(t))g_2(t)$$

and

$$h_3(t) = g_3(t) - 1.$$

It follows that $h_3(t) = -1 < 0$ and $h_3(t) \to +\infty$ as $t \to r_2^-$. Denote the smallest zero of function h_3 by r_3 on the interval $(0, r_2)$. Finally, define the radius of convergence (say, r^*) by

$$r^* = \min\{r_i\}, \quad i = 1, 2, 3. \tag{7}$$

Then, for each $t \in [0, r)$, we have that

$$0 \leq g_i(t) < 1, \quad i = 1, 2, 3. \tag{8}$$

Denote by $U(v, \varepsilon) = \{x \in B_1 : \|x - v\| < \varepsilon\}$ the ball whose center $v \in B_1$ and radius $\varepsilon > 0$. Moreover, $\bar{U}(v, \varepsilon)$ denotes the closure of $U(v, \varepsilon)$.

We will study the local convergence of method (3) in a Banach space setting under the following hypotheses (collectively called by the name 'A'):

(a1) $F : \Omega \subseteq B_1 \to B_2$ is a continuously differentiable operator and $[\cdot, \cdot; F] : \Omega \times \Omega \to \mathcal{L}(B_1, B_2)$ is a first divided difference operator of F.
(a2) There exists $x^* \in \Omega$ so that $F(x^*) = 0$ and $F'(x^*)^{-1} \in \mathcal{L}(B_2, B_1)$.
(a3) There exists a continuous and nondecreasing function $w_0 : \mathbb{R}_+ \cup \{0\} \to \mathbb{R}_+ \cup \{0\}$ with $w_0(0) = 0$ such that, for each $x \in \Omega$,

$$\|F'(x^*)^{-1}([x, y; F] - F'(x^*))\| \leq w_0(\|x - x^*\|, \|y - x^*\|).$$

(a4) Let $\Omega_0 = \Omega \cap U(x^*, r)$, where r has been defined before. There exists continuous and nondecreasing function $v_0 : [0, r) \to \mathbb{R}_+ \cup \{0\}$ such that, for each $x, y \in \Omega_0$,

$$\|\beta[x, x^*; F]\| \leq v_0(\|x_0 - x^*\|),$$

$$\bar{U}(x^*, r) \subset \Omega,$$

$$\|I + \beta[x, x^*; F]\| \leq p.$$

(a5) $\bar{U}(x^*, r_3) \subseteq \Omega$ and $\|F'(x^*)^{-1}F'(x)\| \leq M$.
(a6) Let $R \geq r_3$ and set $\Omega_1 = \Omega \cap \bar{U}(x^*, R)$, $\int_0^1 w_0(\theta R) d\theta < 1$.

Theorem 1. *Suppose that the hypotheses (A) hold. Then, the sequence $\{x_n\}$ generated by method (3) for $x_0 \in U(x^*, r_3) - \{x^*\}$ is well defined in $U(x^*, r_3)$, remains in $U(x^*, r_3)$ and converges to x^*. Moreover, the following conditions hold:*

$$\|y_n - x^*\| \leq g_1(\|x_n - x^*\|)\|x_n - x^*\| \leq \|x_n - x^*\| < \varrho, \tag{9}$$

$$\|z_n - x^*\| \leq g_2(\|x_n - x^*\|)\|x_n - x^*\| \leq \|x_n - x^*\| \tag{10}$$

and

$$\|x_{n+1} - x^*\| \leq g_3(\|x_n - x^*\|)\|x_n - x^*\| \leq \|x_n - x^*\|, \tag{11}$$

where the functions g_i, $i = 1, 2, 3$ are defined as above. Furthermore, the vector x^* is the only solution of $F(x) = 0$ in Ω_1.

Proof. We shall show estimates (9)–(11) using mathematical induction. By hypothesis (a3) and for $x \in U(x^*, r_3)$, we have that

$$\begin{aligned}\|F'(x^*)^{-1}([u_0, x_0; F] - F'(x^*))\| &\leq w_0(\|u_0 - x^*\|, \|x_0 - x^*\|) \\ &\leq w_0(\|x_0 - x^* + \beta F(x_0)\|, \|x_0 - x^*\|) \\ &\leq w_0((I + \beta[x_0, x^*; F])\|x_0 - x^*\|, \|x_0 - x^*\|) \\ &\leq w_0(p\|x_0 - x^*\|, \|x_0 - x^*\|) \\ &\leq w_0(pr, r) < 1.\end{aligned} \tag{12}$$

By (12) and the Banach Lemma [9], we have that $[u_n, x_n; F]^{-1} \in \mathcal{L}(B_2, B_1)$ and

$$\|[u_0, x_0; F]^{-1}F'(x^*)\| \leq \frac{1}{1 - w_0(p\|x_0 - x^*\|, \|x_0 - x^*\|)}. \tag{13}$$

We show that y_n is well defined by the method (3) for $n = 0$. We have

$$\begin{aligned}y_0 - x^* &= x_0 - x^* - [u_0, x_0; F]^{-1}F(x_0) \\ &= [u_0, x_0; F]^{-1}F'(x^*)F'(x^*)^{-1}([u_0, x_0; F] - [x_0, x^*; F])(x_0 - x^*).\end{aligned} \tag{14}$$

Then, using (8) (for $i = 1$), the conditions (a4) and (13), we have in turn that

$$
\begin{aligned}
\|y_0 - x^*\| &= \|[u_0, x_0; F]^{-1} F'(x^*)\| \|F'(x^*)^{-1} ([u_0, x_0; F] - [x_n, x^*; F])\| \|x_0 - x^*\| \\
&\leq \frac{w_1(\|u_0 - x_0\|, \|x_0 - x^*\|) \|x_0 - x^*\|}{1 - w_0(p\|x_0 - x^*\|, \|x_0 - x^*\|)} \\
&\leq \frac{w_1(\|\beta F(x_0)\|, \|x_0 - x^*\|) \|x_0 - x^*\|}{1 - w_0(p\|x_0 - x^*\|, \|x_0 - x^*\|)} \\
&\leq \frac{w_1(\|\beta [x_0, x^*; F](x_0 - x^*)\|, \|x_0 - x^*\|)}{1 - w_0(p\|x_0 - x^*\|, \|x_0 - x^*\|)} \|x_0 - x^*\| \\
&\leq \frac{w_1(\|\beta v_0(\|x_0 - x^*\|)(x_0 - x^*)\|, \|x_0 - x^*\|)}{1 - w_0(p\|x_0 - x^*\|, \|x_0 - x^*\|)} \|x_0 - x^*\| \\
&\leq g_1(\|x_0 - x^*\|) \|x_0 - x^*\| < \|x_0 - x^*\| < r,
\end{aligned}
\tag{15}
$$

which implies (9) for $n = 0$ and $y_0 \in U(x^*, r_3)$.

Note that for each $\theta \in [0, 1]$ and $\|x^* + \theta(x_0 - x^*) - x^*\| = \theta \|x_0 - x^*\| < r$, that is, $x^* + \theta(x_0 - x^*) \in U(x^*, r_3)$, writing

$$
F(x_0) = F(x_0) - F(x^*) = \int_0^1 F'(x^* + \theta(x_0 - x^*))(x_0 - x^*) d\theta. \tag{16}
$$

Then, using (a5), we get that

$$
\begin{aligned}
\|F'(x^*)^{-1} F(x_0)\| &= \left\| \int_0^1 F'(x^*)^{-1} F'(x^* + \theta(x_0 - x^*))(x_0 - x^*) d\theta \right\| \\
&\leq M \|x_0 - x^*\|.
\end{aligned}
\tag{17}
$$

Similarly, we obtain

$$
\|F'(x^*)^{-1} F(y_0)\| \leq M \|y_0 - x^*\|, \tag{18}
$$

$$
\|F'(x^*)^{-1} F(z_0)\| \leq M \|z_0 - x^*\|. \tag{19}
$$

From the second sub-step of method (3), (13), (15) and (18), we obtain that

$$
\begin{aligned}
\|z_0 - x^*\| &\leq \|y_0 - x^*\| + \|[u_0, x_0; F]^{-1} F'(x^*)\| \|F'(x^*)^{-1} F(y_0)\| \\
&\leq \|y_0 - x^*\| + \frac{M\|y_0 - x^*\|}{1 - w_0(p\|x_0 - x^*\|, \|x_0 - x^*\|)} \\
&\leq \left(1 + \frac{M}{1 - w_0(p\|x_0 - x^*\|, \|x_0 - x^*\|)}\right) \|y_0 - x^*\| \\
&\leq \left(1 + \frac{M}{1 - w_0(p\|x_0 - x^*\|, \|x_0 - x^*\|)}\right) g_1(\|x_0 - x^*\|) \|x_0 - x^*\| \\
&\leq g_2(\|x_0 - x^*\|) \|x_0 - x^*\| < \|x_0 - x^*\| < r,
\end{aligned}
\tag{20}
$$

which proves (10) for $n = 0$ and $z_0 \in U(x^*, r_3)$.

Let $\psi(x_n, y_n) = (2I - [u_n, x_n; F]^{-1}[y_n, z_n; F])[u_n, x_n; F]^{-1}$ and notice that, since $x_0, y_0 \in U(x^*, r_3)$, we have that

$$\begin{aligned}
&\|\psi(x_0, y_0)F'(x^*)\| \\
&= \|(2I - [u_0, x_0; F]^{-1}[y_0, z_0; F])[u_0, x_0; F]^{-1}F'(x^*)\| \\
&\leq \left(1 + \|[u_0, x_0; F]^{-1}([u_0, x_0; F] - [y_0, z_0; F])\|\right)\|[u_0, x_0; F]^{-1}F'(x^*)\| \\
&\leq \left(1 + \|[u_0, x_0; F]^{-1}F'(x^*)\|(\|F'(x^*)^{-1}([u_0, x_0; F] - F'(x^*))\| \right. \\
&\quad \left. + \|F'(x^*)^{-1}(F'(x^*) - [y_0, z_0; F])\|)\right)\|[u_0, x_0; F]^{-1}F'(x^*)\| \\
&\leq \left(1 + \frac{\left(w_0(p\|x_n - x^*\|, \|x_n - x^*\|) + w_0(\|y_0 - x^*\|, \|z_0 - x^*\|)\right)}{1 - w_0(p\|x_n - x^*\|, \|x_n - x^*\|)}\right) \quad (21) \\
&\quad \times \frac{1}{1 - w_0(p\|x_n - x^*\|, \|x_n - x^*\|)} \\
&\leq \left(1 + \frac{\left(w_0(p\|x_n - x^*\|, \|x_n - x^*\|) + w_0\left(g_1(\|x_0 - x^*\|)\|x_0 - x^*\|, g_2(\|x_0 - x^*\|)\|x_0 - x^*\|\right)\right)}{1 - w_0(p\|x_n - x^*\|, \|x_n - x^*\|)}\right) \\
&\quad \times \frac{1}{1 - w_0(p\|x_n - x^*\|, \|x_n - x^*\|)} \\
&\leq \bar{g}(\|x_0 - x^*\|).
\end{aligned}$$

Then, using Equation (8) (for $i = 3$), (19), (20) and (21), we obtain

$$\begin{aligned}
\|x_1 - x^*\| &= \|z_0 - x^* - \psi(x_0, y_0)F(z_0)\| \\
&\leq \|z_0 - x^*\| + \|\psi(x_0, y_0)F'(x^*)\|\|F'(x^*)^{-1}F(z_0)\| \\
&\leq \|z_0 - x^*\| + \bar{g}(\|x_0 - x^*\|)M\|z_0 - x^*\| \\
&= (1 + M\bar{g}(\|x_0 - x^*\|))\|z_0 - x^*\| \\
&\leq (1 + M\bar{g}(\|x_0 - x^*\|))g_2(\|x_0 - x^*\|)\|x_0 - x^*\| \\
&\leq g_3(\|x_0 - x^*\|)\|x_0 - x^*\|,
\end{aligned}$$

which proves (11) for $n = 0$ and $x_1 \in U(x^*, r_3)$.

Replace x_0, y_0, z_0, x_1 by x_n, y_n, z_n, x_{n+1} in the preceding estimates to obtain (9)–(11). Then, from the estimates $\|x_{n+1} - x^*\| \leq c\|x_n - x^*\| < r_3$, where $c = g_3(\|x_0 - x^*\|) \in [0, 1)$, we deduce that $\lim_{n \to \infty} x_n = x^*$ and $x_{n+1} \in U(x^*, r_3)$.

Next, we show the uniqueness part using conditions (a3) and (a6). Define operator P by $P = \int_0^1 F'(x^{**} + \theta(x^* - x^{**}))d\theta$ for some $x^{**} \in \Omega_1$ with $F(x^{**}) = 0$. Then, we have that

$$\|F'(x^*)^{-1}(P - F'(x^*))\| \leq \int_0^1 w_0(\theta\|x^* - x^{**}\|)d\theta$$

$$\leq \int_0^1 w_0(\theta\varrho^*)d\theta < 1,$$

so $P^{-1} \in \mathcal{L}(B_2, B_1)$. Then, from the identity

$$0 = F(x^*) - F(x^{**}) = P(x^* - x^{**}),$$

it implies that $x^* = x^{**}$. □

3. Numerical Examples

We illustrate the theoretical results shown in Theorem 1. For the computation of divided difference, let us choose $[x, y; F] = \int_0^1 F'(y + \theta(x - y))d\theta$. Consider the following three numerical examples:

Example 1. *Assume that the motion of a particle in three dimensions is governed by a system of differential equations:*

$$f_1'(x) - f_1(x) - 1 = 0,$$
$$f_2'(y) - (e-1)y - 1 = 0,$$
$$f_3'(z) - 1 = 0,$$

with $x, y, z \in \Omega$ for $f_1(0) = f_2(0) = f_3(0) = 0$. A solution of the system is given for $u = (x, y, z)^T$ by function $F := (f_1, f_2, f_3) : \Omega \to \mathbb{R}^3$ defined by

$$F(u) = \left(e^x - 1, \frac{e-1}{2}y^2 + y, z\right)^T.$$

Its Fréchet-derivative $F'(u)$ is given by

$$F'(u) = \begin{bmatrix} e^x & 0 & 0 \\ 0 & (e-1)y + 1 & 0 \\ 0 & 0 & 1 \end{bmatrix}.$$

Then, for $x^* = (0, 0, 0)^T$, we deduce that $w_0(s, t) = w_1(s, t) = \frac{L_0}{2}(s + t)$ and $v_0(t) = \frac{1}{2}(1 + e^{\frac{1}{L_0}})$, $p = 1 + \frac{1}{2}(1 + e^{\frac{1}{L_0}})$, $\beta = \frac{1}{100}$, where $L_0 = e - 1$ and $M = 2$. Then, using a definition of parameters, the calculated values are displayed as

$$r^* = \min\{r_1, r_2, r_3\} = \min\{0.313084, 0.165881, 0.0715631\} = 0.0715631.$$

Example 2. *Let $X = C[0, 1]$, $\Omega = \bar{U}(x^*, 1)$. We consider the integral equation of the mixed Hammerstein-type [9] given by*

$$x(s) = \int_0^1 k(s, t) \frac{x(t)^2}{2} dt,$$

wherein the kernel k is the green function on the interval $[0, 1] \times [0, 1]$ defined by

$$k(s, t) = \begin{cases} (1 - s)t, & t \leq s, \\ s(1 - t), & s \leq t. \end{cases}$$

Solution $x^*(s) = 0$ is the same as the solution of equation $F(x) = 0$, where $F : C[0, 1]$ is given by

$$F(x)(s) = x(s) - \int_0^1 k(s, t) \frac{x(t)^2}{2} dt.$$

Observe that

$$\left\| \int_0^1 k(s, t) dt \right\| \leq \frac{1}{8}.$$

Then, we have that

$$F'(x)y(s) = y(s) - \int_0^1 k(s, t) x(t) dt,$$

and $F'(x^*(s)) = I$. We can choose $w_0(s,t) = w_1(s,t) = \frac{s+t}{16}$, $v_0(t) = \frac{9}{16}$, $p = \frac{25}{16}$, $\beta = \frac{1}{100}$ and $M = 2$. Then, using a definition of parameters, the calculated values are displayed as

$$r^* = \min\{r_1, r_2, r_3\} = \min\{4.4841, 2.3541, 1.0090\} = 1.0090.$$

Example 3. Let $B_1 = B_2 = C[0,1]$ be the spaces of continuous functions defined on the interval $[0,1]$. Define function F on $\Omega = \bar{U}(0,1)$ by

$$F(\varphi)(x) = \phi(x) - 10 \int_0^1 x\theta\varphi(\theta)^3 d\theta.$$

It follows that

$$F'(\varphi(\xi))(x) = \xi(x) - 30 \int_0^1 x\theta\varphi(\theta)^2 \xi(\theta) d\theta, \text{ for each } \xi \in \Omega.$$

Then, for $x^* = 0$, we have that $w_0(s,t) = w_1(s,t) = L_0(s+t)$ and $v_0(t) = 2$, $p = 3$, $\beta = \frac{1}{100}$, where $L_0 = 15$ and $M = 2$. The parameters are displayed as

$$r^* = \min\{r_1, r_2, r_3\} = \min\{0.013280, 0.0076012, 0.0034654\} = 0.0034654.$$

4. Basins of Attraction

The basin of attraction is a useful geometrical tool for assessing convergence regions of the iterative methods. These basins show us all the starting points that converge to any root when we apply an iterative method, so we can see in a visual way which points are good choices as starting points and which are not. We take the initial point as $z_0 \in R$, where R is a rectangular region in $\mathbb{C}$ containing all the roots of a poynomial $p(z) = 0$. The iterative methods starting at a point z_0 in a rectangle can converge to the zero of the function $p(z)$ or eventually diverge. In order to analyze the basins, we consider the stopping criterion for convergence as 10^{-3} up to a maximum of 25 iterations. If the mentioned tolerance is not attained in 25 iterations, the process is stopped with the conclusion that the iterative method starting at z_0 does not converge to any root. The following strategy is taken into account: A color is assigned to each starting point z_0 in the basin of attraction of a zero. If the iteration starting from the initial point z_0 converges, then it represents the basins of attraction with that particular color assigned to it and, if it fails to converge in 25 iterations, then it shows the black color.

We analyze the basins of attraction on the following two problems:

Test problem 1. Consider the polynomial $p_1(z) = z^4 - 6z^2 + 8$ that has four simple zeros $\{\pm 2, \pm 1.414\ldots\}$. We use a grid of 400×400 points in a rectangle $R \in \mathbb{C}$ of size $[-3,3] \times [-3,3]$ and allocate the red, blue, green and yellow colors to the basins of attraction of these four zeros. Basins obtained for the method (3) are shown in Figure 1(i)–(iii) corresponding to $\beta = 10^{-2}, 10^{-4}, 10^{-9}$. Observing the behavior of the method, we say that the divergent zones (black zones) are becoming smaller with the decreasing value of β.

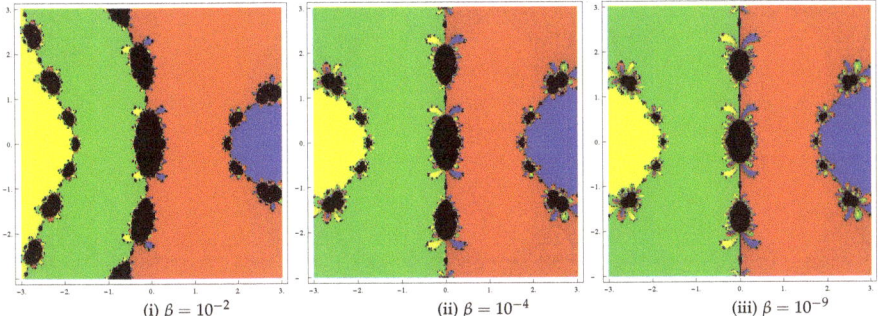

Figure 1. Basins of attraction of method for polynomial $p_1(z)$.

Problem 2. Let us take the polynomial $p_2(z) = z^3 - z$ having zeros $\{0, \pm 1\}$. In this case, we also consider a rectangle $R = [-3, 3] \times [-3, 3] \in \mathbb{C}$ with 400×400 grid points and allocate the colors red, green and blue to each point in the basin of attraction of $-1, 0$ and 1, respectively. Basins obtained for the method (3) are displayed in Figure 2(i)–(iii) for the parameter values $\beta = 10^{-2}$, 10^{-4}, 10^{-9}. Notice that the divergent zones are becoming smaller in size as parameter β assumes smaller values.

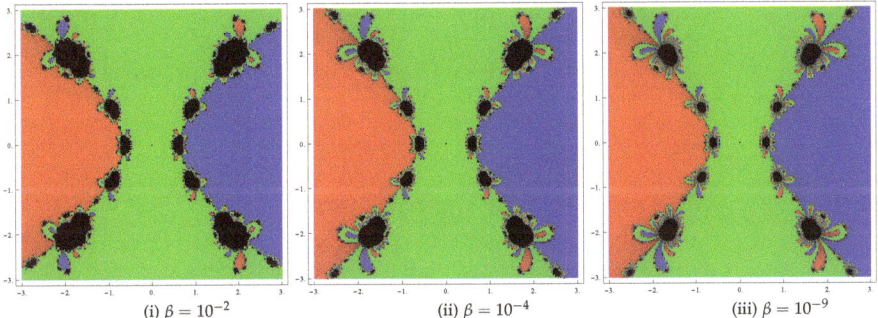

Figure 2. Basins of attraction of method for polynomial $p_2(z)$.

5. Conclusions

In this paper, the local convergence analysis of a derivative-free fifth order method is studied in Banach space. Unlike other techniques that rely on higher derivatives and Taylor series, we have used only derivative of order one in our approach. In this way, we have extended the usage of the considered method since the method can be applied to a wider class of functions. Another advantage of analyzing the local convergence is the computation of a convergence ball, uniqueness of the ball where the iterates lie and estimation of errors. Theoretical results of convergence thus achieved are confirmed through testing on some practical problems.

The basins of attraction have been analyzed by applying the method on some polynomials. From these graphics, one can easily visualize the behavior and suitability of any method. If we choose an initial guess x_0 in a domain where different basins of the roots meet each other, it is uncertain to predict which root is going to be reached by the iterative method that begins from x_0. Thus, the choice of initial guess lying in such a domain is not suitable. In addition, black zones and the zones with different colors are not suitable to take the initial guess x_0 when we want to achieve a particular root. The most attractive pictures appear when we have very intricate boundaries of the basins. Such pictures belong to the cases where the method is more demanding with respect to the initial point.

Author Contributions: Methodology, D.K.; writing, review and editing, J.R.S.; investigation, J.R.S.; data curation, D.K.; conceptualization, L.J.; formal analysis, L.J.

Funding: This research received no external funding.

Acknowledgments: We would like to express our gratitude to the anonymous reviewers for their valuable comments and suggestions which have greatly improved the presentation of this paper.

Conflicts of Interest: The authors declare no conflict of interest.

References

1. Banach, S. *Théorie des Opérations Linéare*; Monografje Matematyczne: Warszawna, Poland, 1932.
2. Gupta, V.; Bora, S.N.; Nieto, J.J. Dhage iterative principle for quadratic perturbation of fractional boundary value problems with finite delay. *Math. Methods Appl. Sci.* **2019**, *42*, 4244–4255. [CrossRef]
3. Jäntschi, L.; Bálint, D.; Bolboacă, S. Multiple linear regressions by maximizing the likelihood under assumption of generalized Gauss-Laplace distribution of the error. *Comput. Math. Methods Med.* **2016**, *2016*, 8578156. [CrossRef] [PubMed]
4. Kitkuan, D.; Kumam, P.; Padcharoen, A.; Kumam, W.; Thounthong, P. Algorithms for zeros of two accretive operators for solving convex minimization problems and its application to image restoration problems. *J. Comput. Appl. Math.* **2019**, *354*, 471–495. [CrossRef]
5. Sachs, M.; Leimkuhler, B.; Danos, V. Langevin dynamics with variable coefficients and nonconservative forces: From stationary states to numerical methods. *Entropy* **2017**, *19*, 647. [CrossRef]
6. Behl, R.; Cordero, A.; Torregrosa, J.R.; Alshomrani, A.S. New iterative methods for solving nonlinear problems with one and several unknowns. *Mathematics* **2018**, *6*, 296. [CrossRef]
7. Argyros, I.K.; George, S. Unified semi-local convergence for k−step iterative methods with flexible and frozen linear operator. *Mathematics* **2018**, *6*, 233. [CrossRef]
8. Argyros, I.K.; Hilout, S. *Computational Methods in Nonlinear Analysis*; World Scientific Publishing Company: Hackensack, NJ, USA, 2013.
9. Argyros, I.K. *Computational Theory of Iterative Methods, Series: Studies in Computational Mathematics 15*; Chui, C.K., Wuytack, L., Eds.; Elsevier: New York, NY, USA, 2007.
10. Traub, J.F. *Iterative Methods for the Solution of Equations*; Prentice-Hall: Englewood Cliffs, NJ, USA, 1964.
11. Potra, F.A.; Ptak, V. *Nondiscrete Induction and Iterative Process*; Research Notes in Mathematics; Pitman: Boston, MA, USA, 1984.
12. Kantrovich, L.V.; Akilov, G.P. *Functional Analysis*; Pergamon Press: Oxford, UK, 1982.
13. Candela, V.; Marquina, A. Recurrence relations for rational cubic methods I: The Halley method. *Computing* **1990**, *44*, 169–184. [CrossRef]
14. Candela, V.; Marquina, A. Recurrence relations for rational cubic methods II: The Chebyshev method. *Computing* **1990**, *45*, 355–367. [CrossRef]
15. Hasanov, V.I.; Ivanov, I.G.; Nebzhibov, G. A new modification of Newton's method. *Appl. Math. Eng.* **2002**, *27*, 278–286.
16. Kou, J.S.; Li, Y.T.; Wang, X.H. A modification of Newton's method with third-order convergence. *Appl. Math. Comput.* **2006**, *181*, 1106–1111. [CrossRef]
17. Ezquerro, J.A.; Hernández, M.A. Recurrence relation for Chebyshev-type methods. *Appl. Math. Optim.* **2000**, *41*, 227–236. [CrossRef]
18. Chun, C.; Stănică, P.; Neta, B. Third-order family of methods in Banach spaces. *Comput. Math. Appl.* **2011**, *61*, 1665–1675. [CrossRef]
19. Hernández, M.A.; Salanova, M.A. Modification of the Kantorovich assumptions for semilocal convergence of the Chebyshev method. *J. Comput. Appl. Math.* **2000**, *126*, 131–143. [CrossRef]
20. Amat, S.; Hernández, M.A.; Romero, N. Semilocal convergence of a sixth order iterative method for quadratic equations. *Appl. Numer. Math.* **2012**, *62*, 833–841. [CrossRef]
21. Babajee, D.K.R.; Dauhoo, M.Z.; Darvishi, M.T.; Barati, A. A note on the local convergence of iterative methods based on Adomian decomposition method and 3-node quadrature rule. *Appl. Math. Comput.* **2008**, *200*, 452–458. [CrossRef]
22. Ren, H.; Wu, Q. Convergence ball and error analysis of a family of iterative methods with cubic convergence. *Appl. Math. Comput.* **2009**, *209*, 369–378. [CrossRef]

23. Ren, H.; Argyros, I.K. Improved local analysis for certain class of iterative methods with cubic convergence. *Numer. Algor.* **2012**, *59*, 505–521. [CrossRef]
24. Argyros, I.K.; Hilout, S. Weaker conditions for the convergence of Newton's method. *J. Complexity* **2012**, *28*, 364–387. [CrossRef]
25. Gutiérrez, J.M.; Magreñán, A.A.; Romero, N. On the semilocal convergence of Newton–Kantorovich method under center-Lipschitz conditions. *Appl. Math. Comput.* **2013**, *221*, 79–88. [CrossRef]
26. Argyros, I.K.; Sharma, J.R.; Kumar, D. Local convergence of Newton–Gauss methods in Banach space. *SeMA* **2016**, *74*, 429–439. [CrossRef]
27. Behl, R.; Salimi, M.; Ferrara, M.; Sharifi, S.; Alharbi, S.K. Some real-life applications of a newly constructed derivative free iterative scheme. *Symmetry* **2019**, *11*, 239. [CrossRef]
28. Salimi, M.; Nik Long, N.M.A.; Sharifi, S.; Pansera, B.A. A multi-point iterative method for solving nonlinear equations with optimal order of convergence. *Jpn. J. Ind. Appl. Math.* **2018**, *35*, 497–509. [CrossRef]
29. Sharma, J.R.; Kumar, D. A fast and efficient composite Newton-Chebyshev method for systems of nonlinear equations. *J. Complexity* **2018**, *49*, 56–73. [CrossRef]
30. Sharma, J.R.; Arora, H. On efficient weighted-Newton methods for solving systems of nonlinear equations. *Appl. Math. Comput.* **2013**, *222*, 497–506. [CrossRef]
31. Lofti, T.; Sharifi, S.; Salimi, M.; Siegmund, S. A new class of three-point methods with optimal convergence order eight and its dynamics. *Numer. Algor.* **2015**, *68*, 261–288. [CrossRef]
32. Sharma, J.R.; Kumar, D.; Jäntschi, L. On a reduced cost higher order Traub–Steffensen-like method for nonlinear systems. *Symmetry* **2019**, *11*, 891. [CrossRef]
33. Grabnier, J.V. Who gave you the epsilon? Cauchy and the origins of rigorous calculus. *Am. Math. Mon.* **1983**, *90*, 185–194. [CrossRef]

© 2019 by the authors. Licensee MDPI, Basel, Switzerland. This article is an open access article distributed under the terms and conditions of the Creative Commons Attribution (CC BY) license (http://creativecommons.org/licenses/by/4.0/).

Article

On Derivative Free Multiple-Root Finders with Optimal Fourth Order Convergence

Janak Raj Sharma [1,*], **Sunil Kumar** [1] **and Lorentz Jäntschi** [2,3,*]

1   Department of Mathematics, Sant Longowal Institute of Engineering and Technology, Longowal Sangrur 148106, India; sfageria1988@gmail.com
2   Department of Physics and Chemistry, Technical University of Cluj-Napoca, 400114 Cluj-Napoca, Romania
3   Institute of Doctoral Studies, Babeş-Bolyai University, 400084 Cluj-Napoca, Romania
*   Correspondence: jrshira@yahoo.co.in (J.R.S.); lorentz.jantschi@gmail.com (L.J.)

Received: 14 June 2020; Accepted: 2 July 2020; Published: 3 July 2020

Abstract: A number of optimal order multiple root techniques that require derivative evaluations in the formulas have been proposed in literature. However, derivative-free optimal techniques for multiple roots are seldom obtained. By considering this factor as motivational, here we present a class of optimal fourth order methods for computing multiple roots without using derivatives in the iteration. The iterative formula consists of two steps in which the first step is a well-known Traub–Steffensen scheme whereas second step is a Traub–Steffensen-like scheme. The Methodology is based on two steps of which the first is Traub–Steffensen iteration and the second is Traub–Steffensen-like iteration. Effectiveness is validated on different problems that shows the robust convergent behavior of the proposed methods. It has been proven that the new derivative-free methods are good competitors to their existing counterparts that need derivative information.

Keywords: multiple root solvers; composite method; weight-function; derivative-free method; optimal convergence

MSC: 65H05; 41A25; 49M15

1. Introduction

Finding root of a nonlinear equation $\psi(u) = 0$ is a very important and interesting problem in many branches of science and engineering. In this work, we examine derivative-free numerical methods to find a multiple root (say, α) with multiplicity μ of the equation $\psi(u) = 0$ that means $\psi^{(j)}(\alpha) = 0, j = 0, 1, 2, ..., \mu - 1$ and $\psi^{(\mu)}(\alpha) \neq 0$. Newton's method [1] is the most widely used basic method for finding multiple roots, which is given by

$$u_{k+1} = u_k - \mu \frac{\psi(u_k)}{\psi'(u_k)}, \quad k = 0, 1, 2, \ldots, \quad \mu = 2, 3, 4, \ldots. \tag{1}$$

A number of modified methods, with or without the base of Newton's method, have been elaborated and analyzed in literature, see [2–14]. These methods use derivatives of either first order or both first and second order in the iterative scheme. Contrary to this, higher order methods without derivatives to calculate multiple roots are yet to be examined. These methods are very useful in the problems where the derivative ψ' is cumbersome to evaluate or is costly to compute. The derivative-free counterpart of classical Newton method (1) is the Traub–Steffensen method [15]. The method uses the approximation

$$\psi'(u_k) \simeq \frac{\psi(u_k + \beta \psi(u_k)) - \psi(u_k)}{\beta \psi(u_k)}, \quad \beta \in \mathbb{R} - \{0\},$$

or
$$\psi'(u_k) \simeq \psi[v_k, u_k],$$

for the derivative ψ' in the Newton method (1). Here, $v_k = u_k + \beta \psi(u_k)$ and $\psi[v_k, u_k] = \frac{\psi(v_k) - \psi(u_k)}{v_k - u_k}$ is a first order divided difference. Thereby, the method (1) takes the form of the Traub–Steffensen scheme defined as

$$u_{k+1} = u_k - \mu \frac{\psi(u_k)}{\psi[v_k, u_k]}. \qquad (2)$$

The Traub–Steffensen method (2) is a prominent improvement of the Newton method because it maintains the quadratic convergence without adding any derivative.

Unlike Newton-like methods, the Traub–Steffensen-like methods are difficult to construct. Recently, a family of two-step Traub–Steffensen-like methods with fourth order convergence has been proposed in [16]. In terms of computational cost, the methods of [16] use three function evaluations per iteration and thus possess optimal fourth order convergence according to Kung–Traub conjecture (see [17]). This hypothesis states that multi-point methods without memory requiring m functional evaluations can attain the convergence order 2^{m-1} called optimal order. Such methods are usually known as optimal methods. Our aim in this work is to develop derivative-free multiple root methods of good computational efficiency, which is to say, the methods of higher convergence order with the amount of computational work as small as we please. Consequently, we introduce a class of Traub–Steffensen-like derivative-free fourth order methods that require three new pieces of information of the function ψ and therefore have optimal fourth order convergence according to Kung–Traub conjecture. The iterative formula consists of two steps with Traub–Steffensen iteration (2) in the first step, whereas there is Traub–Steffensen-like iteration in the second step. Performance is tested numerically on many problems of different kinds. Moreover, comparison of performance with existing modified Newton-like methods verifies the robust and efficient nature of the proposed methods.

We summarize the contents of paper. In Section 2, the scheme of fourth order iteration is formulated and convergence order is studied separately for different cases. The main result, showing the unification of different cases, is studied in Section 3. Section 4 contains the basins of attractors drawn to assess the convergence domains of new methods. In Section 5, numerical experiments are performed on different problems to demonstrate accuracy and efficiency of the methods. Concluding remarks about the work are reported in Section 6.

2. Development of a Novel Scheme

Researchers have used different approaches to develop higher order iterative methods for solving nonlinear equations. Some of them are: Interpolation approach, Sampling approach, Composition approach, Geometrical approach, Adomian approach, and Weight-function approach. Of these, the Weight-function approach has been most popular in recent times; see, for example, Refs. [10,13,14,18,19] and references therein. Using this approach, we consider the following two-step iterative scheme for finding multiple root with multiplicity $\mu \geq 2$:

$$\begin{aligned} z_k &= u_k - \mu \frac{\psi(u_k)}{\psi[v_k, u_k]}, \\ u_{k+1} &= z_k - G(h)\left(1 + \frac{1}{y_k}\right)\frac{\psi(u_k)}{\psi[v_k, u_k]}, \end{aligned} \qquad (3)$$

where $h = \frac{x_k}{1+x_k}$, $x_k = \sqrt[\mu]{\frac{\psi(z_k)}{\psi(u_k)}}$, $y_k = \sqrt[\mu]{\frac{\psi(v_k)}{\psi(u_k)}}$ and $G : \mathbb{C} \to \mathbb{C}$ is analytic in the neighborhood of zero. This iterative scheme is weighted by the factors $G(h)$ and $\left(1 + \frac{1}{y_k}\right)$, hence the name weight-factor or weight-function technique.

Note that x_k and y_k are one-to-μ multi-valued functions, so we consider their principal analytic branches [18]. Hence, it is convenient to treat them as the principal root. For example, let us consider the case of x_k. The principal root is given by $x_k = \exp\left[\frac{1}{\mu}\text{Log}\left(\frac{\psi(z_k)}{\psi(u_k)}\right)\right]$, with $\text{Log}\left(\frac{\psi(z_k)}{\psi(u_k)}\right) = \text{Log}\left|\frac{\psi(z_k)}{\psi(u_k)}\right| + i\,\text{Arg}\left(\frac{\psi(z_k)}{\psi(u_k)}\right)$ for $-\pi < \text{Arg}\left(\frac{\psi(z_k)}{\psi(u_k)}\right) \leq \pi$; this convention of $\text{Arg}(p)$ for $p \in \mathbb{C}$ agrees with that of $\text{Log}[p]$ command of Mathematica [20] to be employed later in the sections of basins of attraction and numerical experiments. Similarly, we treat for y_k.

In the sequel, we prove fourth order of convergence of the proposed iterative scheme (3). For simplicity, the results are obtained separately for the cases depending upon the multiplicity μ. Firstly, we consider the case $\mu = 2$.

Theorem 1. *Assume that $u = \alpha$ is a zero with multiplicity $\mu = 2$ of the function $\psi(u)$, where $\psi : \mathbb{C} \to \mathbb{C}$ is sufficiently differentiable in a domain containing α. Suppose that the initial point u_0 is closer to α; then, the order of convergence of the scheme (3) is at least four, provided that the weight-function $G(h)$ satisfies the conditions $G(0) = 0$, $G'(0) = 1$, $G''(0) = 6$ and $|G'''(0)| < \infty$.*

Proof. Assume that $\varepsilon_k = u_k - \alpha$ is the error at the k-th stage. Expanding $\psi(u_k)$ about α using the Taylor series keeping in mind that $\psi(\alpha) = 0$, $\psi'(\alpha) = 0$ and $\psi^{(2)}(\alpha) \neq 0$, we have that

$$\psi(u_k) = \frac{\psi^{(2)}(\alpha)}{2!}\varepsilon_k^2\left(1 + A_1\varepsilon_k + A_2\varepsilon_k^2 + A_3\varepsilon_k^3 + A_4\varepsilon_k^4 + \cdots\right), \tag{4}$$

where $A_m = \frac{2!}{(2+m)!}\frac{\psi^{(2+m)}(\alpha)}{\psi^{(2)}(\alpha)}$ for $m \in \mathbb{N}$.

Similarly, Taylor series expansion of $\psi(v_k)$ is

$$\psi(v_k) = \frac{\psi^{(2)}(\alpha)}{2!}\varepsilon_{v_k}^2\left(1 + A_1\varepsilon_{v_k} + A_2\varepsilon_{v_k}^2 + A_3\varepsilon_{v_k}^3 + A_4\varepsilon_{v_k}^4 + \cdots\right), \tag{5}$$

where $\varepsilon_{v_k} = v_k - \alpha = \varepsilon_k + \frac{\beta\psi^{(2)}(\alpha)}{2!}\varepsilon_k^2\left(1 + A_1\varepsilon_k + A_2\varepsilon_k^2 + A_3\varepsilon_k^3 + A_4\varepsilon_k^4 + \cdots\right)$.

By using (4) and (5) in the first step of (3), we obtain

$$\varepsilon_{z_k} = z_k - \alpha$$
$$= \frac{1}{2}\left(\frac{\beta\psi^{(2)}(\alpha)}{2} + A_1\right)\varepsilon_k^2 - \frac{1}{16}\left((\beta\psi^{(2)}(\alpha))^2 - 8\beta\psi^{(2)}(\alpha)A_1 + 12A_1^2 - 16A_2\right)\varepsilon_k^3 + \frac{1}{64}\left((\beta\psi^{(2)}(\alpha))^3\right. \tag{6}$$
$$\left. - 20\beta\psi^{(2)}(\alpha)A_1^2 + 72A_1^3 + 64\beta\psi^{(2)}(\alpha)A_2 - 10A_1((\beta\psi^{(2)}(\alpha))^2 + 16A_2) + 96A_3\right)\varepsilon_k^4 + O(\varepsilon_k^5).$$

In addition, we have that

$$\psi(z_k) = \frac{\psi^{(2)}(\alpha)}{2!}\varepsilon_{z_k}^2\left(1 + A_1\varepsilon_{z_k} + A_2\varepsilon_{z_k}^2 + \cdots\right). \tag{7}$$

Using (4), (5) and (7), we further obtain

$$x_k = \frac{1}{2}\left(\frac{\beta\psi^{(2)}(\alpha)}{2} + A_1\right)\varepsilon_k - \frac{1}{16}\left((\beta\psi^{(2)}(\alpha))^2 - 6\beta\psi^{(2)}(\alpha)A_1 + 16(A_1^2 - A_2)\right)\varepsilon_k^2 + \frac{1}{64}\left((\beta\psi^{(2)}(\alpha))^3\right.$$
$$\left. - 22\beta\psi^{(2)}(\alpha)A_1^2 + 4(29A_1^3 + 14\beta\psi^{(2)}(\alpha)A_2) - 2A_1(3(\beta\psi^{(2)}(\alpha))^2 + 104A_2) + 96A_3\right)\varepsilon_k^3 \tag{8}$$
$$+ \frac{1}{256}\left(212\beta\psi^{(2)}(\alpha)A_1^3 - 800A_1^4 + 2A_1^2(-7(\beta\psi^{(2)}(\alpha))^2 + 1040A_2) + 2A_1(3(\beta\psi^{(2)}(\alpha))^3 - 232\beta\psi^{(2)}(\alpha)A_2\right.$$
$$\left. - 576A_3) - ((\beta\psi^{(2)}(\alpha))^4 + 8\beta\psi^{(2)}(\alpha)A_2 + 640A_2^2 - 416\beta\psi^{(2)}(\alpha)A_3 - 512A_4)\right)\varepsilon_k^4 + O(\varepsilon_k^5)$$

and

$$y_k = 1 + \frac{\beta\psi^{(2)}(\alpha)}{2}\varepsilon_k\left(1 + \frac{3}{2}A_1\varepsilon_k + \frac{1}{4}(\beta\psi^{(2)}(\alpha)A_1 + 8A_2)\varepsilon_k^2 + \frac{1}{16}(3\beta\psi^{(2)}(\alpha)A_1^2 + 12\beta\psi^{(2)}(\alpha)A_2 + 40A_3)\varepsilon_k^3 \right.$$
$$\left. + O(\varepsilon_k^4)\right). \tag{9}$$

Using (8), we have

$$h = \frac{1}{2}\left(\frac{\beta\psi^{(2)}(\alpha)}{2} + A_1\right)\varepsilon_k - \frac{1}{8}((\beta\psi^{(2)}(\alpha))^2 - \beta\psi^{(2)}(\alpha)A_1 - 2(4A_2 - 5A_1^2))\varepsilon_k^2 + \frac{1}{32}(-\beta\psi^{(2)}(\alpha)A_1^2 + 94A_1^3$$
$$- 4A_1((\beta\psi^{(2)}(\alpha))^2 + 34A_2) + 2((\beta\psi^{(2)}(\alpha))^3 + 6\beta\psi^{(2)}(\alpha)A_2 + 24A_3))\varepsilon_k^3 + \frac{1}{128}(54\beta\psi^{(2)}(\alpha)A_1^3 - 864A_1^4 \quad (10)$$
$$+ A_1^2(1808A_2 - 13(\beta\psi^{(2)}(\alpha))^2) + 2A_1(6(\beta\psi^{(2)}(\alpha))^3 - 68\beta\psi^{(2)}(\alpha)A_2 - 384A_3) - 4((\beta\psi^{(2)}(\alpha))^4$$
$$+ 5(\beta\psi^{(2)}(\alpha))^2 A_2 + 112A_2^2 - 28\beta\psi^{(2)}(\alpha)A_3 - 64A_4))\varepsilon_k^4 + O(\varepsilon_k^5).$$

Taylor expansion of the weight function $G(h)$ in the neighborhood of origin up to third-order terms is given by

$$G(h) \approx G(0) + hG'(0) + \frac{1}{2}h^2 G''(0) + \frac{1}{6}h^2 G'''(0). \quad (11)$$

Using (4)–(11) in the last step of (3), we have

$$\varepsilon_{k+1} = -G(0)\varepsilon_k + \frac{1}{4}(\beta\psi^{(2)}(\alpha)(1 + 2G(0) - G'(0)) + 2(1 + G(0) - G'(0))A_1)\varepsilon_k^2 + \sum_{n=1}^{2}\phi_n \varepsilon_k^{n+2} + O(\varepsilon_k^5), \quad (12)$$

where $\phi_n = \phi_n(\beta, A_1, A_2, A_3, G(0), G'(0), G''(0), G'''(0))$, $n = 1, 2$. The expressions of ϕ_1 and ϕ_2 being very lengthy have not been produced explicitly.

We can obtain at least fourth order convergence if we set coefficients of ε_k, ε_k^2 and ε_k^3 simultaneously equal to zero. Then, some simple calculations yield

$$G(0) = 0, \quad G'(0) = 1, \quad G''(0) = 6. \quad (13)$$

Using (13) in (12), we will obtain final error equation

$$\varepsilon_{k+1} = -\frac{1}{192}\left(\frac{\beta\psi^{(2)}(\alpha)}{2} + A_1\right)\left((G'''(0) - 42)(\beta\psi^{(2)}(\alpha))^2 + 4(G'''(0) - 45)\beta\psi^{(2)}(\alpha)A_1 + 4(G'''(0) - 63)A_1^2 \quad (14)$$
$$+ 48A_2\right)\varepsilon_k^4 + O(\varepsilon_k^5).$$

Thus, the theorem is proved. □

Next, we prove the following theorem for case $\mu = 3$.

Theorem 2. *Using assumptions of Theorem 1, the convergence order of scheme (3) for the case $\mu = 3$ is at least 4, if $G(0) = 0$, $G'(0) = \frac{3}{2}$, $G''(0) = 9$ and $|G'''(0)| < \infty$.*

Proof. Taking into account that $\psi(\alpha) = 0$, $\psi'(\alpha) = 0$, $\psi''(\alpha) = 0$ and $\psi^{(3)}(\alpha) \neq 0$, the Taylor series development of $\psi(u_k)$ about α gives

$$\psi(u_k) = \frac{\psi^{(3)}(\alpha)}{3!}\varepsilon_k^3(1 + B_1\varepsilon_k + B_2\varepsilon_k^2 + B_3\varepsilon_k^3 + B_4\varepsilon_k^4 + \cdots), \quad (15)$$

where $B_m = \frac{3!}{(3+m)!}\frac{\psi^{(3+m)}(\alpha)}{\psi^{(3)}(\alpha)}$ for $m \in \mathbb{N}$.

Expanding $\psi(v_k)$ about α

$$\psi(v_k) = \frac{\psi^{(3)}(\alpha)}{3!}\varepsilon_{v_k}^3(1 + B_1\varepsilon_{v_k} + B_2\varepsilon_{v_k}^2 + B_3\varepsilon_{v_k}^3 + B_4\varepsilon_{v_k}^4 + \cdots), \quad (16)$$

where $\varepsilon_{v_k} = v_k - \alpha = \varepsilon_k + \frac{\beta\psi^{(3)}(\alpha)}{3!}\varepsilon_k^3(1 + B_1\varepsilon_k + B_2\varepsilon_k^2 + B_3\varepsilon_k^3 + B_4\varepsilon_k^4 + \cdots).$

Then, using (15) and (16) in the first step of (3), we obtain

$$\varepsilon_{z_k} = z_k - \alpha$$
$$= \frac{B_1}{3}\varepsilon_k^2 + \frac{1}{18}(3\beta\psi^{(3)}(\alpha) - 8B_1^2 + 12B_2)\varepsilon_k^3 + \frac{1}{27}(16B_1^3 + 3B_1(2\beta\psi^{(3)}(\alpha) - 13B_2) + 27B_3)\varepsilon_k^4 + O(\varepsilon_k^5). \quad (17)$$

Expansion of $\psi(z_k)$ about α yields

$$\psi(z_k) = \frac{\psi^{(3)}(\alpha)}{3!}\varepsilon_{z_k}^3(1 + B_1\varepsilon_{z_k} + B_2\varepsilon_{z_k}^2 + B_3\varepsilon_{z_k}^3 + B_4\varepsilon_{z_k}^4 + \cdots). \quad (18)$$

Then, from (15), (16), and (18), it follows that

$$x_k = \frac{B_1}{3}\varepsilon_k + \frac{1}{18}(3\beta\psi^{(3)}(\alpha) - 10B_1^2 + 12B_2)\varepsilon_k^2 + \frac{1}{54}\left(46B_1^3 + 3B_1(3\psi^{(3)}(\alpha)\beta - 32B_2) + 54B_3\right)\varepsilon_k^3 - \frac{1}{486}(610B_1^4 \quad (19)$$
$$- B_1^2(1818B_2 - 27\beta\psi^{(3)}(\alpha)) + 1188B_1B_3 + 9((\beta\psi^{(3)}(\alpha))^2 - 15\beta\psi^{(3)}(\alpha)B_2 + 72B_2^2 - 72B_4))\varepsilon_k^4 + O(\varepsilon_k^5)$$

and

$$y_k = 1 + \frac{\beta\psi^{(3)}(\alpha)}{3!}\varepsilon_k^2\left(1 + \frac{4}{3}B_1\varepsilon_k + \frac{5}{3}B_2\varepsilon_k^2 + \frac{1}{18}(\beta\psi^{(3)}(\alpha)B_1 + 36B_3)\varepsilon_k^3 + O(\varepsilon_k^4)\right). \quad (20)$$

Using (19), we have

$$h = \frac{B_1}{3}\varepsilon_k + \frac{1}{6}(\beta\psi^{(3)}(\alpha) - 4B_1^2 + 4B_2)\varepsilon_k^2 + \frac{1}{54}\left(68B_1^3 + 3B_1(\beta\psi^{(3)}(\alpha) - 40B_2) + 54B_3\right)\varepsilon_k^3 - \frac{1}{2916}(6792B_1^4$$
$$- 108B_1^2(159B_2 + 2\beta\psi^{(3)}(\alpha)) + 9072B_1B_3 - 27(-5(\beta\psi^{(3)}(\alpha))^2 + 6\beta\psi^{(3)}(\alpha)B_2 - 192B_2^2 + 144B_4))\varepsilon_k^4 \quad (21)$$
$$+ O(\varepsilon_k^5).$$

Developing weight function $G(h)$ about origin by the Taylor series expansion,

$$G(h) \approx G(0) + hG'(0) + \frac{1}{2}h^2G''(0) + \frac{1}{6}h^3G'''(0). \quad (22)$$

By using (15)–(22) in the last step of (3), we have

$$\varepsilon_{k+1} = -\frac{2G(0)}{3}\varepsilon_k + \frac{1}{9}(3 + 2G(0) - 2G'(0))B_1\varepsilon_k^2 + \sum_{n=1}^{2}\varphi_n\varepsilon_k^{n+2} + O(\varepsilon_k^5), \quad (23)$$

where $\varphi_n = \varphi_n(\beta, B_1, B_2, B_3, G(0), G'(0), G''(0), G'''(0))$, $n = 1, 2$.

To obtain fourth order convergence, it is sufficient to set coefficients of ε_k, ε_k^2, and ε_k^3 simultaneously equal to zero. This process will yield

$$G(0) = 0, \quad G'(0) = \frac{3}{2}, \quad G''(0) = 9. \quad (24)$$

Then, error equation (23) is given by

$$\varepsilon_{k+1} = -\frac{B_1}{972}(27\beta\psi^{(3)}(\alpha) + 4(G'''(0) - 99)B_1^2 + 108B_2)\varepsilon_k^4 + O(\varepsilon_k^5). \quad (25)$$

Hence, the result is proved. □

Remark 1. *We can observe from the above results that the number of conditions on $G(h)$ is 3 corresponding to the cases $\mu = 2, 3$ to attain the fourth order convergence of the method (3). These cases also satisfy common conditions: $G(0) = 0$, $G'(0) = \frac{\mu}{2}$, $G''(0) = 3\mu$. Their error equations also contain the term involving the parameter β. However, for the cases $\mu \geq 4$, it has been seen that the error equation in each such case does not contain β term. We shall prove this fact in the next section.*

3. Main Result

We shall prove the convergence order of scheme (3) for the multiplicity $\mu \geq 4$ by the following theorem:

Theorem 3. *Using assumptions of Theorem 1, the convergence order of scheme (3) for $\mu \geq 4$ is at least four, provided that $G(0) = 0$, $G'(0) = \frac{\mu}{2}$, $G''(0) = 3\mu$ and $|G'''(0)| < \infty$. Moreover, error in the scheme is given by*

$$\varepsilon_{k+1} = \frac{1}{6\mu^4}\left((3\mu(19+\mu) - 2G'''(0))F_1^3 - 6\mu^2 F_1 F_2\right)\varepsilon_k^4 + O(\varepsilon_k^5),$$

where $F_m = \frac{\mu!}{(\mu+m)!} \frac{\psi^{(\mu+m)}(\alpha)}{\psi^{(\mu)}(\alpha)}$ for $m \in \mathbb{N}$.

Proof. Taking into account that $\psi^{(i)}(\alpha) = 0$, $i = 0, 1, 2, \ldots, \mu - 1$ and $\psi^{(\mu)}(\alpha) \neq 0$, then Taylor series expansion of $\psi(u_k)$ about α is

$$\psi(u_k) = \frac{\psi^{(\mu)}(\alpha)}{\mu!}\varepsilon_k^\mu\left(1 + F_1\varepsilon_k + F_2\varepsilon_k^2 + F_3\varepsilon_k^3 + F_4\varepsilon_k^4 + \cdots\right). \tag{26}$$

Taylor expansion of $\psi(v_k)$ about α yields

$$\psi(v_k) = \frac{\psi^{(\mu)}(\alpha)}{\mu!}\varepsilon_{v_k}^\mu\left(1 + F_1\varepsilon_{v_k} + F_2\varepsilon_{v_k}^2 + F_3\varepsilon_{v_k}^3 + F_4\varepsilon_{v_k}^4 + \cdots\right), \tag{27}$$

where $\varepsilon_{v_k} = v_k - \alpha = \varepsilon_k + \frac{\beta \psi^{(\mu)}(\alpha)}{\mu!}\varepsilon_k^\mu\left(1 + F_1\varepsilon_k + F_2\varepsilon_k^2 + F_3\varepsilon_k^3 + F_4\varepsilon_k^4 + \cdots\right)$.

Using (26) and (27) in the first step of (3), we obtain

$$\varepsilon_{z_k} = \begin{cases} \frac{F_1}{4}\varepsilon_k^2 + \frac{1}{16}(8F_2 - 5F_1^2)\varepsilon_k^3 + \frac{1}{64}(4\beta\psi^{(4)}(\alpha) + 25F_1^3 - 64F_1F_2 + 48F_3)\varepsilon_k^4 + O(\varepsilon_k^5), & \text{if } \mu = 4, \\ \frac{F_1}{\mu}\varepsilon_k^2 + \frac{1}{\mu^2}(2\mu F_2 - (1+\mu)F_1^2)\varepsilon_k^3 + \frac{1}{\mu^3}((1+\mu)^2 F_1^3 - \mu(4+3\mu)F_1F_2 + 3\mu^2 F_3)\varepsilon_k^4 + O(\varepsilon_k^5), & \text{if } \mu \geq 5, \end{cases} \tag{28}$$

where $\varepsilon_{z_k} = z_k - \alpha$.

Expansion of $\psi(z_k)$ around α yields

$$\psi(z_k) = \frac{\psi^{(\mu)}(\alpha)}{\mu!}\varepsilon_{z_k}^\mu\left(1 + F_1\varepsilon_{z_k} + F_2\varepsilon_{z_k}^2 + F_3\varepsilon_{z_k}^3 + F_4\varepsilon_{z_k}^4 + \cdots\right). \tag{29}$$

Using (26), (27) and (29), we have that

$$x_k = \begin{cases} \frac{F_1}{4}\varepsilon_k + \frac{1}{8}(4F_2 - 3F_1^2)\varepsilon_k^2 + \frac{1}{128}(8\beta\psi^{(4)}(\alpha) + 67F_1^3 - 152F_1F_2 + 96F_3)\varepsilon_k^3 + \frac{1}{768}\big(-543F_1^4 + 1740F_1^2 F_2 \\ \quad + 4F_1(11\beta\psi^{(4)}(\alpha) - 312F_3) + 96(-7F_2^2 + 8F_4)\big)\varepsilon_k^4 + O(\varepsilon_k^5), & \text{if } \mu = 4, \\ \frac{F_1}{5}\varepsilon_k + \frac{1}{25}(10F_2 - 7F_1^2)\varepsilon_k^2 + \frac{1}{125}(46F_1^3 - 110F_1F_2 + 75F_3)\varepsilon_k^3 + \big(\frac{\beta\psi^{(5)}(\alpha)}{60} - \frac{294}{625}F_1^4 + \frac{197}{125}F_1^2 F_2 - \frac{16}{25}F_2^2 \\ \quad - \frac{6}{5}F_1 F_3 + \frac{4}{5}F_4\big)\varepsilon_k^4 + O(\varepsilon_k^5), & \text{if } \mu = 5, \\ \frac{F_1}{\mu}\varepsilon_k + \frac{1}{\mu^2}(2\mu F_2 - (2+\mu)F_1^2)\varepsilon_k^2 + \frac{1}{2\mu^3}((7 + 7\mu + 2\mu^2)F_1^3 - 2\mu(7 + 3\mu)F_1F_2 + 6\mu^2 F_3)\varepsilon_k^3 \\ \quad - \frac{1}{6\mu^4}\big((34 + 51\mu + 29\mu^2 + 6\mu^3)F_1^4 - 6\mu(17 + 16\mu + 4\mu^2)F_1^2 F_2 + 12\mu^2(3+\mu)F_2^2 \\ \quad + 12\mu^2(5 + 2\mu)F_1 F_3\big)\varepsilon_k^4 + O(\varepsilon_k^5), & \text{if } \mu \geq 6 \end{cases} \tag{30}$$

and

$$y_k = 1 + \frac{\beta\psi^{(\mu)}(\alpha)}{\mu!}\varepsilon_k^{\mu-1}\left(1 + \frac{(\mu+1)F_1}{\mu}\varepsilon_k + \frac{(\mu+2)F_2}{\mu}\varepsilon_k^2 + \frac{(\mu+3)F_3}{\mu}\varepsilon_k^3 + \frac{(\mu+4)F_4}{\mu}\varepsilon_k^4 + O(\varepsilon_k^5)\right). \tag{31}$$

Using (30), we obtain that

$$h = \begin{cases} \frac{F_1}{4}\varepsilon_k + \frac{1}{16}(8\mu F_2 - 7F_1^2)\varepsilon_k^2 + \frac{1}{128}(8\beta\psi^{(4)}(\alpha) + 93F_1^3 - 184F_1F_2 + 96F_3)\varepsilon_k^3 \\ \quad + (-\frac{303}{256}F_1^4 + \frac{213}{64}F_1^2F_2 - \frac{9}{8}F_2^2 + F_1(\frac{5}{192}\beta\psi^{(4)}(\alpha) - 2F_3) + F_4)\varepsilon_k^4 + O(\varepsilon_k^5), \text{ if } \mu = 4, \\ \frac{F_1}{5}\varepsilon_k + \frac{1}{25}(10F_2 - 8F_1^2)\varepsilon_k^2 + \frac{1}{125}(61F_1^3 - 130F_1F_2 + 75F_3)\varepsilon_k^3 \\ \quad + (-\frac{457}{625}F_1^4 + \frac{11}{5}F_1^2F_2 - \frac{36}{25}F_1F_3 + \frac{1}{60}(\beta\psi^{(5)}(\alpha) - 48F_2^2 + 48F_4))\varepsilon_k^4 + O(\varepsilon_k^5), \text{ if } \mu = 5, \\ \frac{F_1}{\mu}\varepsilon_k + \frac{1}{\mu^2}(2\mu F_2 - (3+\mu)F_1^2)\varepsilon_k^2 + \frac{1}{2\mu^3}((17+11\mu+2\mu^2)F_1^3 - 2\mu(11+3\mu)F_1F_2 + 6\mu^2F_3)\varepsilon_k^3 \\ \quad - \frac{1}{6\mu^4}((142+135\mu+47\mu^2+6\mu^3)F_1^4 - 6\mu(45+26\mu+4\mu^2)F_1^2F_2 + 12\mu^2(5+\mu)F_2^2 \\ \quad + 24\mu^2(4+\mu)F_1F_3)\varepsilon_k^4 + O(\varepsilon_k^5), \text{ if } \mu \geq 6. \end{cases} \quad (32)$$

Developing weight function $G(h)$ about origin by the Taylor series expansion,

$$G(h) \approx G(0) + hG'(0) + \frac{1}{2}h^2G''(0) + \frac{1}{6}h^3G'''(0). \quad (33)$$

Using (26)–(33) in the last step of (3), we get

$$\varepsilon_{k+1} = -\frac{2G(0)}{\mu}\varepsilon_k + \frac{1}{\mu^2}((2G(0) - 2G'(0) + \mu)F_1)\varepsilon_k^2 + \sum_{n=1}^{2}\chi_n\varepsilon_k^{n+2} + O(\varepsilon_k^5), \quad (34)$$

where $\chi_n = \chi_n(\beta, F_1, F_2, F_3, G(0), G'(0), G''(0), G'''(0))$ when $\mu = 4, 5$ and $\chi_n = \chi_n(F_1, F_2, F_3, G(0), G'(0), G''(0), G'''(0))$ when $\mu \geq 6$ for $n = 1, 2$.

The fourth order convergence can be attained if we put coefficients of ε_k, ε_k^2 and ε_k^3 simultaneously equal to zero. Then, the resulting equations yield

$$G(0) = 0, \quad G'(0) = \frac{\mu}{2}, \quad G''(0) = 3\mu. \quad (35)$$

As a result, the error equation is given by

$$\varepsilon_{k+1} = \frac{1}{6\mu^4}((3\mu(19+\mu) - 2G'''(0))F_1^3 - 6\mu^2F_1F_2)\varepsilon_k^4 + O(\varepsilon_k^5). \quad (36)$$

This proves the result. □

Remark 2. *The proposed scheme (3) achieves fourth-order convergence with the conditions of weight-function $G(h)$ as shown in Theorems 1–3. This convergence rate is attained by using only three functional evaluations viz. $\psi(u_k)$, $\psi(v_k)$ and $\psi(z_k)$ per iteration. Therefore, the iterative scheme (3) is optimal according to Kung–Traub conjecture [17].*

Remark 3. *Note that the parameter β, which is used in v_k, appears only in the error equations of the cases $\mu = 2, 3$ but not for $\mu \geq 4$ (see Equation (36)). However, for $\mu \geq 4$, we have observed that this parameter appears in the terms of ε_k^5 and higher order. Such terms are difficult to compute in general. However, we do not need these in order to show the required fourth order of convergence. Note also that Theorems 1–3 are presented to show the difference in error expressions. Nevertheless, the weight function $G(h)$ satisfies the common conditions $G(0) = 0, G'(0) = \frac{\mu}{2}, G''(0) = 3\mu$ for every $\mu \geq 2$.*

Some Special Cases

Based on various forms of function $G(h)$ that satisfy the conditions of Theorem 3, numerous special cases of the family (3) can be explored. The following are some simple forms:

(1) $G(h) = \dfrac{\mu h(1+3h)}{2}$, (2) $G(h) = \dfrac{\mu h}{2-6h}$, (3) $G(h) = \dfrac{\mu h(\mu - 2h)}{2(\mu - (2+3\mu)h + 2\mu h^2)}$,

(4) $G(h) = \dfrac{\mu h(3-h)}{6-20h}$.

The corresponding method to each of the above forms can be expressed as follows:

Method 1 (M1):
$$u_{k+1} = z_k - \dfrac{\mu h(1+3h)}{2}\left(1 + \dfrac{1}{y_k}\right)\dfrac{\psi(u_k)}{\psi[v_k, u_k]}.$$

Method 2 (M2):
$$u_{k+1} = z_k - \dfrac{\mu h}{2-6h}\left(1 + \dfrac{1}{y_k}\right)\dfrac{\psi(u_k)}{\psi[v_k, u_k]}.$$

Method 3 (M3):
$$u_{k+1} = z_k - \dfrac{\mu h(\mu - 2h)}{2(\mu - (2+3\mu)h + 2\mu h^2)}\left(1 + \dfrac{1}{y_k}\right)\dfrac{\psi(u_k)}{\psi[v_k, u_k]}.$$

Method 4 (M4):
$$u_{k+1} = z_k - \dfrac{\mu h(3-h)}{6-20h}\left(1 + \dfrac{1}{y_k}\right)\dfrac{\psi(u_k)}{\psi[v_k, u_k]}.$$

Note that, in all the above cases, z_k has the following form:
$$z_k = u_k - \mu \dfrac{\psi(u_k)}{\psi[v_k, u_k]}.$$

4. Basins of Attraction

In this section, we present complex geometry of the above considered method with a tool, namely basin of attraction, by applying the method to some complex polynomials $\psi(z)$. Basin of attraction of the root is an important geometrical tool for comparing convergence regions of the iterative methods [21–23]. To start with, let us recall some basic ideas concerned with this graphical tool.

Let $R : \mathbb{C} \to \mathbb{C}$ be a rational mapping on the Riemann sphere. We define orbit of a point $z_0 \in \mathbb{C}$ as the set $\{z_0, R(z_0), R^2(z_0), \ldots, R^n(z_0), \ldots\}$. A point $z_0 \in \mathbb{C}$ is a fixed point of the rational function R if it satisfies the equation $R(z_0) = z_0$. A point z_0 is said to be periodic with period $m > 1$ if $R^m(z_0) = z_0$, where m is the smallest such integer. A point z_0 is called attracting if $|R'(z_0)| < 1$, repelling if $|R'(z_0)| > 1$, neutral if $|R'(z_0)| = 1$ and super attracting if $|R'(z_0)| = 0$. Assume that z_ψ^* is an attracting fixed point of the rational map R. Then, the basin of attraction of z_ψ^* is defined as

$$A(z_\psi^*) = \{z_0 \in \mathbb{C} : R^n(z_0) \to z_\psi^*, n \to \infty\}.$$

The set of points whose orbits tend to an attracting fixed point z_ψ^* is called the Fatou set. The complementary set, called the Julia set, is the closure of the set of repelling fixed points, which establishes the boundaries between the basins of the roots. Attraction basins allow us to assess those starting points which converge to the concerned root of a polynomial when we apply an iterative method, so we can visualize which points are good options as starting points and which are not.

We select z_0 as the initial point belonging to D, where D is a rectangular region in $\mathbb{C}$ containing all the roots of the equation $\psi(z) = 0$. An iterative method starting with a point $z_0 \in D$ may converge to the zero of the function $\psi(z)$ or may diverge. To assess the basins, we consider 10^{-3} as the stopping criterion for convergence restricted to 25 iterations. If this tolerance is not achieved in the required iterations, the procedure is dismissed with the result showing the divergence of the iteration function started from z_0. While drawing the basins, the following criterion is adopted: A color is allotted to every initial guess z_0 in the attraction basin of a zero. If the iterative formula that begins at point z_0 converges, then it forms the basins of attraction with that assigned color and, if the formula fails to converge in the required number of iterations, then it is painted black.

To view the complex dynamics, the proposed methods are applied on the following three problems:

Test problem 1. Consider the polynomial $\psi_1(z) = (z^2 + z + 1)^2$ having two zeros $\{-0.5 - 0.866025i, -0.5 + 0.866025i\}$ with multiplicity $\mu = 2$. The attraction basins for this polynomial are shown in Figures 1–3 corresponding to the choices 0.01, 10^{-4}, 10^{-6} of parameter β. A color is assigned to each basin of attraction of a zero. In particular, red and green colors have been allocated to the basins of attraction of the zeros $-0.5 - 0.866025i$ and $-0.5 + 0.866025i$, respectively.

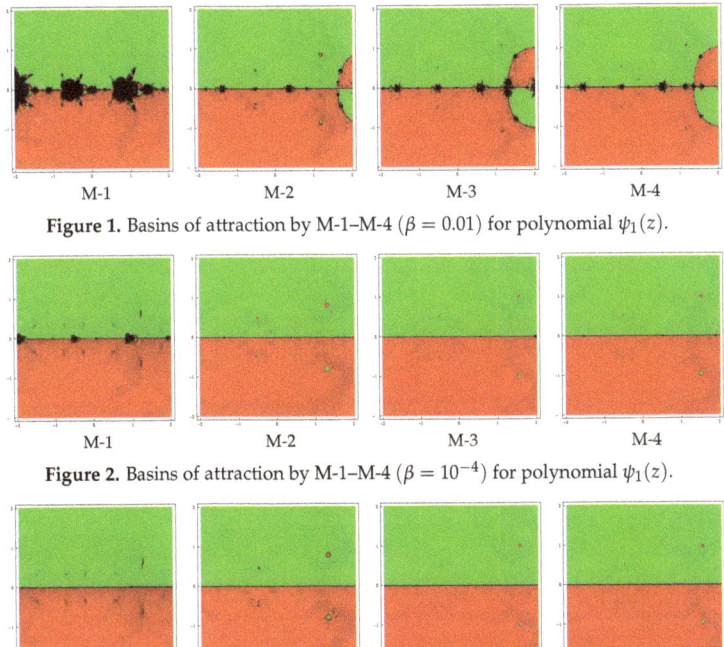

Figure 1. Basins of attraction by M-1–M-4 ($\beta = 0.01$) for polynomial $\psi_1(z)$.

Figure 2. Basins of attraction by M-1–M-4 ($\beta = 10^{-4}$) for polynomial $\psi_1(z)$.

Figure 3. Basins of attraction by M-1–M-4 ($\beta = 10^{-6}$) for polynomial $\psi_1(z)$.

Test problem 2. Consider the polynomial $\psi_2(z) = \left(z^3 + \frac{1}{4}z\right)^3$ which has three zeros $\{-\frac{i}{2}, \frac{i}{2}, 0\}$ with multiplicities $\mu = 3$. Basins of attractors assessed by methods for this polynomial are drawn in Figures 4–6 corresponding to choices $\beta = 0.01$, 10^{-4}, 10^{-6}. The corresponding basin of a zero is identified by a color assigned to it. For example, green, red, and blue colors have been assigned corresponding to $-\frac{i}{2}, \frac{i}{2}$, and 0.

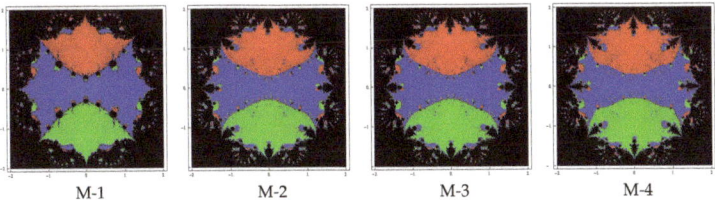

Figure 4. Basins of attraction by M-1–M-4 ($\beta = 0.01$) for polynomial $\psi_2(z)$.

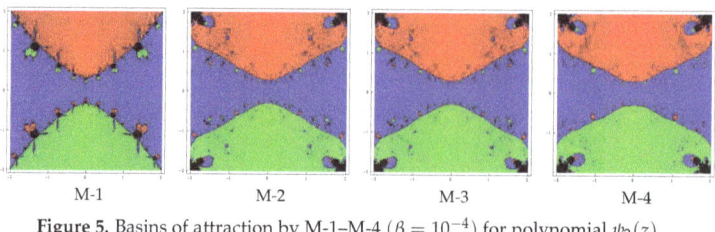

Figure 5. Basins of attraction by M-1–M-4 ($\beta = 10^{-4}$) for polynomial $\psi_2(z)$.

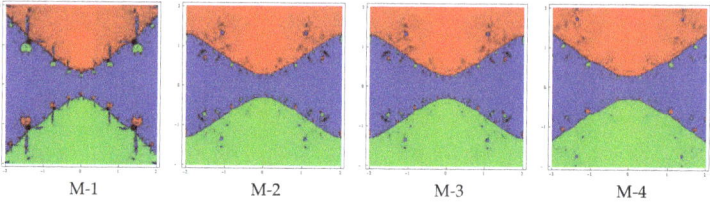

Figure 6. Basins of attraction by methods M-1–M-4 ($\beta = 10^{-6}$) for polynomial $\psi_2(z)$.

Test problem 3. Next, let us consider the polynomial $\psi_3(z) = \left(z^3 + \frac{1}{2}\right)^4$ that has four zeros $\{-0.707107 + 0.707107i, -0.707107 - 0.707107i, 0.707107 + 0.707107i, 0.707107 - 0.707107i\}$ with multiplicity $\mu = 4$. The basins of attractors of zeros are shown in Figures 7–9, for choices of the parameter $\beta = 0.01$, 10^{-4}, 10^{-6}. A color is assigned to each basin of attraction of a zero. In particular, we assign yellow, blue, red, and green colors to $-0.707107 + 0.707107i$, $-0.707107 - 0.707107i$, $0.707107 + 0.707107i$ and $0.707107 - 0.707107i$, respectively.

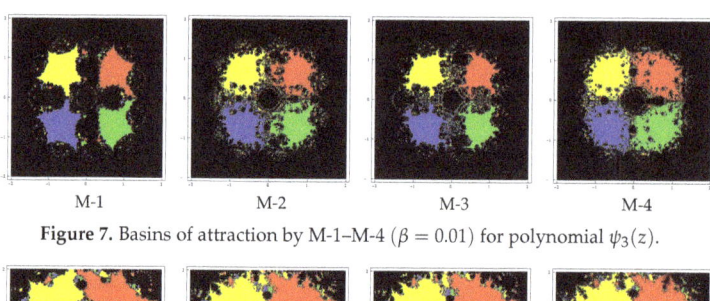

Figure 7. Basins of attraction by M-1–M-4 ($\beta = 0.01$) for polynomial $\psi_3(z)$.

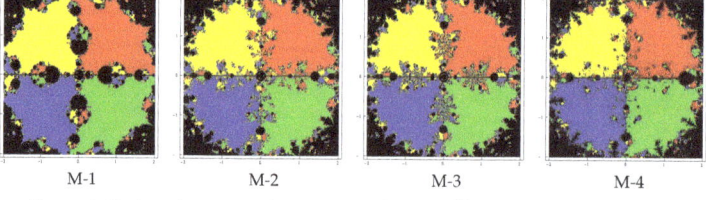

Figure 8. Basins of attraction by M-1–M-4 ($\beta = 10^{-4}$) for polynomial $\psi_3(z)$.

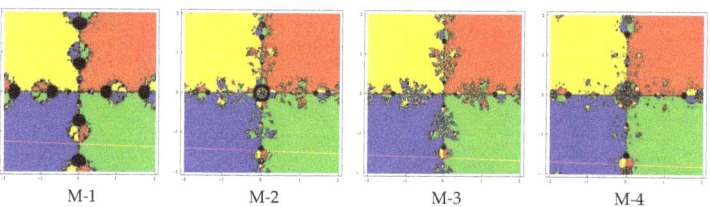

Figure 9. Basins of attraction by M-1–M-4 ($\beta = 10^{-6}$) for polynomial $\psi_3(z)$.

Estimation of β values plays an important role in the selection of those members of family (3) which possess good convergence behavior. This is also the reason why different values of β have been chosen to assess the basins. The above graphics clearly indicate that basins are becoming wider with the smaller values of parameter β. Moreover, the black zones (used to indicate divergence zones) are also diminishing as β assumes small values. Thus, we conclude this section with a remark that the convergence of proposed methods is better for smaller values of parameter β.

5. Numerical Results

In order to validate of theoretical results that have been shown in previous sections, the new methods M1, M2, M3, and M4 are tested numerically by implementing them on some nonlinear equations. Moreover, these are compared with some existing optimal fourth order Newton-like methods. For example, we consider the methods by Li–Liao–Cheng [7], Li–Cheng–Neta [8], Sharma–Sharma [9], Zhou–Chen–Song [10], Soleymani–Babajee–Lotfi [12], and Kansal–Kanwar–Bhatia [14]. The methods are expressed as follows:

Li–Liao–Cheng method (LLCM):

$$z_k = u_k - \frac{2\mu}{\mu+2} \frac{\psi(u_k)}{\psi'(u_k)},$$

$$u_{k+1} = u_k - \frac{\mu(\mu-2)\left(\frac{\mu}{\mu+2}\right)^{-\mu}\psi'(z_k) - \mu^2 \psi'(u_k)}{\psi'(u_k) - \left(\frac{\mu}{\mu+2}\right)^{-\mu}\psi'(z_k)} \frac{\psi(u_k)}{2\psi'(u_k)}.$$

Li–Cheng–Neta method (LCNM):

$$z_k = u_k - \frac{2\mu}{\mu+2} \frac{\psi(u_k)}{\psi'(u_k)},$$

$$u_{k+1} = u_k - \alpha_1 \frac{\psi(u_k)}{\psi'(z_k)} - \frac{\psi(u_k)}{\alpha_2 \psi'(u_k) + \alpha_3 \psi'(z_k)},$$

where

$$\alpha_1 = -\frac{1}{2}\frac{\left(\frac{\mu}{\mu+2}\right)^\mu \mu(\mu^4 + 4\mu^3 - 16\mu - 16)}{\mu^3 - 4\mu + 8},$$

$$\alpha_2 = -\frac{(\mu^3 - 4\mu + 8)^2}{\mu(\mu^4 + 4\mu^3 - 4\mu^2 - 16\mu + 16)(\mu^2 + 2\mu - 4)},$$

$$\alpha_3 = \frac{\mu^2(\mu^3 - 4\mu + 8)}{\left(\frac{\mu}{\mu+2}\right)^\mu (\mu^4 + 4\mu^3 - 4\mu^2 - 16\mu + 16)(\mu^2 + 2\mu - 4)}.$$

Sharma–Sharma method (SSM):

$$z_k = u_k - \frac{2\mu}{\mu+2}\frac{\psi(u_k)}{\psi'(u_k)},$$

$$u_{k+1} = u_k - \frac{\mu}{8}\bigg[(\mu^3 - 4\mu + 8) - (\mu+2)^2\left(\frac{\mu}{\mu+2}\right)^\mu \frac{\psi'(u_k)}{\psi'(z_k)}$$

$$\times \left(2(\mu-1) - (\mu+2)\left(\frac{\mu}{\mu+2}\right)^\mu \frac{\psi'(u_k)}{\psi'(z_k)}\right)\bigg]\frac{\psi(u_k)}{\psi'(u_k)}.$$

Zhou–Chen–Song method (ZCSM):

$$z_k = u_k - \frac{2\mu}{\mu+2}\frac{\psi(u_k)}{\psi'(u_k)},$$

$$u_{k+1} = u_k - \frac{\mu}{8}\left[\mu^3\left(\frac{\mu+2}{\mu}\right)^{2\mu}\left(\frac{\psi'(z_k)}{\psi'(u_k)}\right)^2 - 2\mu^2(\mu+3)\left(\frac{\mu+2}{\mu}\right)^\mu\frac{\psi'(z_k)}{\psi'(u_k)}\right.$$
$$\left. + (\mu^3+6\mu^2+8\mu+8)\right]\frac{\psi(u_k)}{\psi'(u_k)}.$$

Soleymani–Babajee–Lotfi method (SBLM):

$$z_k = u_k - \frac{2\mu}{\mu+2}\frac{\psi(u_k)}{\psi'(u_k)},$$

$$u_{k+1} = u_k - \frac{\psi'(z_k)\psi(u_k)}{q_1(\psi'(z_k))^2 + q_2\psi'(z_k)\psi'(u_k) + q_3(\psi'(u_k))^2},$$

where $q_1 = \frac{1}{16}\mu^{3-\mu}(\mu+2)^\mu$, $q_2 = \frac{8-\mu(\mu+2)(\mu^2-2)}{8\mu}$, $q_3 = \frac{1}{16}(\mu-2)\mu^{\mu-1}(\mu+2)^{3-\mu}$.

Kansal–Kanwar–Bhatia method (KKBM):

$$z_k = u_k - \frac{2\mu}{\mu+2}\frac{\psi(u_k)}{\psi'(u_k)},$$

$$u_{k+1} = u_k - \frac{\mu}{4}\psi(u_k)\left(1 + \frac{\mu^4 p^{-2\mu}\left(p^{\mu-1} - \frac{\psi'(z_k)}{\psi'(u_k)}\right)^2(p^\mu-1)}{8(2p^\mu + n(p^\mu-1))}\right)$$
$$\times\left(\frac{4-2\mu+\mu^2(p^{-\mu}-1)}{\psi'(u_k)} - \frac{p^{-\mu}(2p^\mu+\mu(p^\mu-1))^2}{\psi'(u_k)-\psi'(z_k)}\right),$$

where $p = \frac{\mu}{\mu+2}$.

Computations are performed in the programming package of Mathematica software [20] in a PC with specifications: Intel(R) Pentium(R) CPU B960 @ 2.20 GHz, 2.20 GHz (32-bit Operating System) Microsoft Windows 7 Professional and 4 GB RAM. Numerical tests are performed by choosing the value −0.01 for parameter β in new methods. The tabulated results of the methods displayed in Table 1 include: (i) iteration number (k) required to obtain the desired solution satisfying the condition $|u_{k+1} - u_k| + |\psi(u_k)| < 10^{-100}$, (ii) estimated error $|u_{k+1} - u_k|$ in the consecutive first three iterations, (iii) calculated convergence order (CCO), and (iv) time consumed (CPU time in seconds) in execution of a program, which is measured by the command "TimeUsed[ ]". The calculated convergence order (CCO) is computed by the well-known formula (see [24])

$$\text{CCO} = \frac{\log|(u_{k+2}-\alpha)/(u_{k+1}-\alpha)|}{\log|(u_{k+1}-\alpha)/(u_k-\alpha)|}, \quad \text{for each } k=1,2,\ldots \tag{37}$$

Table 1. Comparison of numerical results.

Methods	k	$|u_2 - u_1|$	$|u_3 - u_2|$	$|u_4 - u_3|$	CCO	CPU-Time
$\psi_1(u)$						
LLCM	6	7.84×10^{-2}	6.31×10^{-3}	1.06×10^{-5}	4.000	0.0784
LCNM	6	7.84×10^{-2}	6.31×10^{-3}	1.06×10^{-5}	4.000	0.0822
SSM	6	7.99×10^{-2}	6.78×10^{-3}	1.44×10^{-5}	4.000	0.0943
ZCSM	6	8.31×10^{-2}	7.83×10^{-3}	2.76×10^{-5}	4.000	0.0956
SBLM	6	7.84×10^{-2}	6.31×10^{-3}	1.06×10^{-5}	4.000	0.0874
KKBM	6	7.74×10^{-2}	5.97×10^{-3}	7.31×10^{-6}	4.000	0.0945
M1	6	9.20×10^{-2}	1.16×10^{-2}	1.16×10^{-4}	4.000	0.0774
M2	6	6.90×10^{-2}	3.84×10^{-3}	1.03×10^{-6}	4.000	0.0794
M3	6	6.21×10^{-2}	2.39×10^{-3}	7.06×10^{-8}	4.000	0.0626
M4	6	6.29×10^{-2}	2.54×10^{-3}	9.28×10^{-8}	4.000	0.0785
$\psi_2(u)$						
LLCM	4	2.02×10^{-4}	2.11×10^{-17}	2.51×10^{-69}	4.000	0.7334
LCNM	4	2.02×10^{-4}	2.12×10^{-17}	2.54×10^{-69}	4.000	1.0774
SSM	4	2.02×10^{-4}	2.12×10^{-17}	2.60×10^{-69}	4.000	1.0765
ZCSM	4	2.02×10^{-4}	2.15×10^{-17}	2.75×10^{-69}	4.000	1.1082
SBLM	4	2.02×10^{-4}	2.13×10^{-17}	2.62×10^{-69}	4.000	1.2950
KKBM	4	2.02×10^{-4}	2.08×10^{-17}	2.31×10^{-69}	4.000	1.1548
M1	4	1.01×10^{-4}	1.08×10^{-18}	1.43×10^{-74}	4.000	0.5612
M2	4	9.85×10^{-5}	4.94×10^{-19}	3.13×10^{-76}	4.000	0.5154
M3	4	9.85×10^{-5}	4.94×10^{-19}	3.13×10^{-76}	4.000	0.5311
M4	4	9.82×10^{-5}	4.35×10^{-19}	1.67×10^{-76}	4.000	0.5003
$\psi_3(u)$						
LLCM	4	4.91×10^{-5}	5.70×10^{-21}	1.03×10^{-84}	4.000	0.6704
LCNM	4	4.91×10^{-5}	5.70×10^{-21}	1.03×10^{-84}	4.000	0.9832
SSM	4	4.92×10^{-5}	5.71×10^{-21}	1.04×10^{-84}	4.000	1.0303
ZCSM	4	4.92×10^{-5}	5.72×10^{-21}	1.05×10^{-84}	4.000	1.0617
SBLM	4	4.92×10^{-5}	5.73×10^{-21}	1.06×10^{-84}	4.000	1.2644
KKBM	4	4.91×10^{-5}	5.66×10^{-21}	1.00×10^{-84}	4.000	1.0768
M1	3	6.35×10^{-6}	2.73×10^{-25}	0	4.000	0.3433
M2	3	4.94×10^{-6}	6.81×10^{-26}	0	4.000	0.2965
M3	3	5.02×10^{-6}	7.46×10^{-26}	0	4.000	0.3598
M4	3	4.77×10^{-6}	5.66×10^{-26}	0	4.000	0.3446
$\psi_4(u)$						
LLCM	4	1.15×10^{-4}	5.69×10^{-17}	3.39×10^{-66}	4.000	1.4824
LCNM	4	1.15×10^{-4}	5.70×10^{-17}	3.40×10^{-66}	4.000	2.5745
SSM	4	1.15×10^{-4}	5.71×10^{-17}	3.44×10^{-66}	4.000	2.5126
ZCSM	4	1.15×10^{-4}	5.72×10^{-17}	3.47×10^{-66}	4.000	2.5587
SBLM	4	1.15×10^{-4}	5.83×10^{-17}	3.79×10^{-66}	4.000	3.1824
KKBM	4	1.15×10^{-4}	5.63×10^{-17}	3.21×10^{-66}	4.000	2.4965
M1	4	4.18×10^{-4}	6.03×10^{-19}	2.60×10^{-74}	4.000	0.4993
M2	4	3.88×10^{-5}	2.24×10^{-19}	2.45×10^{-76}	4.000	0.5151
M3	4	3.92×10^{-5}	2.57×10^{-19}	4.80×10^{-76}	4.000	0.4996
M4	4	3.85×10^{-5}	1.92×10^{-19}	1.18×10^{-76}	4.000	0.4686
$\psi_5(u)$						
LLCM	4	2.16×10^{-4}	3.17×10^{-17}	1.48×10^{-68}	4.000	1.9042
LCNM	4	2.16×10^{-4}	3.17×10^{-17}	1.47×10^{-68}	4.000	2.0594
SSM	4	2.16×10^{-4}	3.16×10^{-17}	1.45×10^{-68}	4.000	2.0125
ZCSM	4	2.16×10^{-4}	3.15×10^{-17}	1.43×10^{-68}	4.000	2.1530
SBLM	4	2.16×10^{-4}	3.01×10^{-17}	1.15×10^{-68}	4.000	2.4185
KKBM	4	2.16×10^{-4}	3.24×10^{-17}	1.63×10^{-68}	4.000	2.2153
M1	4	2.48×10^{-4}	7.62×10^{-21}	6.81×10^{-83}	4.000	1.6697
M2	4	2.15×10^{-5}	2.03×10^{-21}	1.63×10^{-85}	4.000	1.7793
M3	4	2.19×10^{-5}	2.51×10^{-21}	4.35×10^{-85}	4.000	1.7942
M4	4	2.11×10^{-5}	1.66×10^{-21}	6.29×10^{-86}	4.000	1.6855

The problems considered for numerical testing are shown in Table 2.

Table 2. Test functions.

Functions	Root (α)	Multiplicity	Initial Guess
$\psi_1(u) = u^3 - 5.22u^2 + 9.0825u - 5.2675$	1.75	2	2.4
$\psi_2(u) = -\frac{u^4}{12} + \frac{u^2}{2} + u + e^u(u-3) + \sin u + 3$	0	3	0.6
$\psi_3(u) = \left(e^{-u} - 1 + \frac{u}{5}\right)^4$	4.9651142317...	4	5.5
$\psi_4(u) = u(u^2+1)(2e^{u^2+1} + u^2 - 1)\cosh^4\left(\frac{\pi u}{2}\right)$	i	6	1.2 i
$\psi_5(u) = \left[\tan^{-1}\left(\frac{\sqrt{5}}{2}\right) - \tan^{-1}(\sqrt{u^2-1}) + \sqrt{6}\left(\tan^{-1}\left(\sqrt{\frac{u^2-1}{6}}\right)\right.\right.$ $\left.\left. - \tan^{-1}\left(\frac{1}{2}\sqrt{\frac{5}{6}}\right)\right) - \frac{11}{63}\right]^7$	1.8411294068...	7	1.6

From the computed results in Table 1, we can observe the good convergence behavior of the proposed methods. The reason for good convergence is the increase in accuracy of the successive approximations as is evident from values of the differences $|u_{k+1} - u_k|$. This also implies to stable nature of the methods. Moreover, the approximations to solutions computed by the proposed methods have either greater or equal accuracy than those computed by existing counterparts. The value 0 of $|u_{k+1} - u_k|$ indicates that the stopping criterion $|u_{k+1} - u_k| + |\psi(u_k)| < 10^{-100}$ has been satisfied at this stage. From the calculation of calculated convergence order as shown in the second last column in each table, we have verified the theoretical fourth order of convergence. The robustness of new algorithms can also be judged by the fact that the used CPU time is less than that of the CPU time by the existing techniques. This conclusion is also confirmed by similar numerical experiments on many other different problems.

6. Conclusions

We have proposed a family of fourth order derivative-free numerical methods for obtaining multiple roots of nonlinear equations. Analysis of the convergence has been carried out under standard assumptions, which proves the convergence order four. The important feature of our designed scheme is its optimal order of convergence which is rare to achieve in derivative-free methods. Some special cases of the family have been explored. These cases are employed to solve some nonlinear equations. The performance is compared with existing techniques of a similar nature. Testing of the numerical results have shown the presented derivative-free method as good competitors to the already established optimal fourth order techniques that use derivative information in the algorithm. We conclude this work with a remark: the proposed derivative-free methods can be a better alternative to existing Newton-type methods when derivatives are costly to evaluate.

Author Contributions: Methodology, J.R.S.; Writing—review & editing, J.R.S.; Investigation, S.K.; Data Curation, S.K.; Conceptualization, L.J.; Formal analysis, L.J. All authors have read and agreed to the published version of the manuscript.

Funding: This research received no external funding.

Conflicts of Interest: The authors declare no conflict of interest.

References

1. Schröder, E. Über unendlich viele Algorithmen zur Auflösung der Gleichungen. *Math. Ann.* **1870**, *2*, 317–365. [CrossRef]
2. Hansen E.; Patrick, M. A family of root finding methods. *Numer. Math.* **1977**, *27*, 257–269. [CrossRef]
3. Victory, H.D.; Neta, B. A higher order method for multiple zeros of nonlinear functions. *Int. J. Comput. Math.* **1983**, *12*, 329–335. [CrossRef]
4. Dong, C. A family of multipoint iterative functions for finding multiple roots of equations. *Int. J. Comput. Math.* **1987**, *21*, 363–367. [CrossRef]
5. Osada, N. An optimal multiple root-finding method of order three. *J. Comput. Appl. Math.* **1994**, *51*, 131–133. [CrossRef]

6. Neta, B. New third order nonlinear solvers for multiple roots. *App. Math. Comput.* **2008**, *202*, 162–170. [CrossRef]
7. Li, S.; Liao, X.; Cheng, L. A new fourth-order iterative method for finding multiple roots of nonlinear equations. *Appl. Math. Comput.* **2009**, *215*, 1288–1292.
8. Li, S.G.; Cheng, L.Z.; Neta, B. Some fourth-order nonlinear solvers with closed formulae for multiple roots. *Comput Math. Appl.* **2010**, *59*, 126–135. [CrossRef]
9. Sharma, J.R.; Sharma, R. Modified Jarratt method for computing multiple roots. *Appl. Math. Comput.* **2010**, *217*, 878–881. [CrossRef]
10. Zhou, X.; Chen, X.; Song, Y. Constructing higher-order methods for obtaining the multiple roots of nonlinear equations. *J. Comput. Appl. Math.* **2011**, *235*, 4199–4206. [CrossRef]
11. Sharifi, M.; Babajee, D.K.R.; Soleymani, F. Finding the solution of nonlinear equations by a class of optimal methods. *Comput. Math. Appl.* **2012**, *63*, 764–774. [CrossRef]
12. Soleymani, F.; Babajee, D.K.R.; Lotfi, T. On a numerical technique for finding multiple zeros and its dynamics. *J. Egypt. Math. Soc.* **2013**, *21*, 346–353. [CrossRef]
13. Geum, Y.H.; Kim Y.I.; Neta, B. A class of two-point sixth-order multiple-zero finders of modified double-Newton type and their dynamics. *Appl. Math. Comput.* **2015**, *270*, 387–400. [CrossRef]
14. Kansal, M.; Kanwar, V.; Bhatia, S. On some optimal multiple root-finding methods and their dynamics. *Appl. Appl. Math.* **2015**, *10*, 349–367.
15. Traub, J.F. *Iterative Methods for the Solution of Equations*; Chelsea Publishing Company: New York, NY, USA, 1982.
16. Sharma, J.R.; Kumar, S.; Jäntschi, L. On a class of optimal fourth order multiple root solvers without using derivatives. *Symmetry* **2019**, *11*, 766. [CrossRef]
17. Kung, H.T.; Traub, J.F. Optimal order of one-point and multipoint iteration. *J. Assoc. Comput. Mach.* **1974**, *21*, 643–651. [CrossRef]
18. Geum, Y.H.; Kim, Y.I.; Neta, B. Constructing a family of optimal eighth-order modified Newton-type multiple-zero finders along with the dynamics behind their purely imaginary extraneous fixed points. *J. Comp. Appl. Math.* **2018**, *333*, 131–156. [CrossRef]
19. Benbernou, S.; Gala, S.; Ragusa, M.A. On the regularity criteria for the 3D magnetohydrodynamic equations via two components in terms of BMO space. *Math. Meth. Appl. Sci.* **2016**, *37*, 2320–2325. [CrossRef]
20. Wolfram, S. *The Mathematica Book*, 5th ed.; Wolfram Media: Champaign, IL, USA, 2003.
21. Vrscay, E.R.; Gilbert, W.J. Extraneous fixed points, basin boundaries and chaotic dynamics for Schröder and König rational iteration functions. *Numer. Math.* **1988**, *52*, 1–16. [CrossRef]
22. Varona, J.L. Graphic and numerical comparison between iterative methods. *Math. Intell.* **2002**, *24*, 37–46. [CrossRef]
23. Argyros, I.K.; Magreñán, Á.A. *Iterative Methods and Their Dynamics with Applications*; CRC Press: New York, NY, USA, 2017.
24. Weerakoon, S.; Fernando, T.G.I. A variant of Newton's method with accelerated third-order convergence. *Appl. Math. Lett.* **2000**, *13*, 87–93. [CrossRef]

© 2020 by the authors. Licensee MDPI, Basel, Switzerland. This article is an open access article distributed under the terms and conditions of the Creative Commons Attribution (CC BY) license (http://creativecommons.org/licenses/by/4.0/).

Article

Finite Integration Method with Shifted Chebyshev Polynomials for Solving Time-Fractional Burgers' Equations

Ampol Duangpan [1], Ratinan Boonklurb [1,*] and Tawikan Treeyaprasert [2]

[1] Department of Mathematics and Computer Science, Faculty of Science, Chulalongkorn University, Bangkok 10330, Thailand; ty_math@hotmail.com
[2] Department of Mathematics and Statistics, Faculty of Science, Thammasat University, Rangsit Center, Pathum Thani 12120, Thailand; tawikan@tu.ac.th
* Correspondence: ratinan.b@chula.ac.th

Received: 21 October 2019; Accepted: 3 December 2019; Published: 7 December 2019

Abstract: The Burgers' equation is one of the nonlinear partial differential equations that has been studied by many researchers, especially, in terms of the fractional derivatives. In this article, the numerical algorithms are invented to obtain the approximate solutions of time-fractional Burgers' equations both in one and two dimensions as well as time-fractional coupled Burgers' equations which their fractional derivatives are described in the Caputo sense. These proposed algorithms are constructed by applying the finite integration method combined with the shifted Chebyshev polynomials to deal the spatial discretizations and further using the forward difference quotient to handle the temporal discretizations. Moreover, numerical examples demonstrate the ability of the proposed method to produce the decent approximate solutions in terms of accuracy. The rate of convergence and computational cost for each example are also presented.

Keywords: finite integration method; shifted Chebyshev polynomial; Caputo fractional derivative; Burgers' equation; coupled Burgers' equation

---

1. Introduction

Fractional calculus has received much attention due to the fact that several real-world phenomena can be demonstrated successfully by developing mathematical models using fractional calculus. More specifically, fractional differential equations (FDEs) are the generalized form of integer order differential equations. The applications of the FDEs have been emerging in many fields of science and engineering such as diffusion processes [1], thermal conductivity [2], oscillating dynamical systems [3], rheological models [4], quantum models [5], etc. However, one of the interesting issues for the FDEs is a fractional Burgers' equation. It appears in many areas of applied mathematics and can describe various kinds of phenomena such as mathematical models of turbulence and shock wave traveling, formation, and decay of nonplanar shock waves at the velocity fluctuation of sound, physical processes of unidirectional propagation of weakly nonlinear acoustic waves through a gas-filled pipe, and so on, see [6–8]. In order to understand these phenomena as well as further apply them in the practical life, it is important to find their solutions. Some powerful numerical methods had been developed for solving the fractional Burgers' equation, such as finite difference methods (FDM) [9], Adomian decomposition method [10], and finite volume method [11]. Moreover, in 2015, Esen and Tasbozan [12] gave a numerical solution of time fractional Burgers' equation by assuming that the solution $u(x,t)$ can be approximated by a linear combination of products of two functions, one of which involves only x and the other involves only t. Recently, Yokus and kaya [13] used the FDM to find the numerical solution for time fractional Burgers' equation, however, their results contained less accuracy. In 2017,

Cao et al. [14] studied solution of two-dimensional time-fractional Burgers' equation with high and low Reynolds numbers using discontinuous Galerkin method, however, the method involves the triangulations of the domain which usually gives difficulty in terms of devising a computational program. There are more numerical studies on time- and/or space-fractional Burgers' equations which can be found in many researches.

In this article, we present the numerical technique based on the finite integration method (FIM) for solving time-fractional Burger' equations and time-fractional coupled Burgers' equations. The FIM is one of the interesting numerical methods in solving partial differential equations (PDEs). The idea of using FIM is to transform the given PDE into an equivalent integral equation and apply numerical integrations to solve the integral equation afterwards. It is known that the numerical integration is very insensitive to round-off errors, while numerical differentiation is very sensitive to round-off errors. It is because the manipulation task of numerical differentiation involves division by small step-size but the process of numerical integration involves multiplication by small step-size.

Originally, the FIM has been firstly proposed by Wen et al. [15]. They constructed the integration matrices based on trapezoidal rule and radial basis functions for solving one-dimensional linear PDEs and then Li et al. [16] continued to develop it in order to overcome the two-dimensional problems. After that, the FIM was improved using three numerical quadratures, including Simpson's rule, Newton-Cotes, and Lagrange interpolation, presented by Li et al. [17]. The FIM has been successfully applied to solve various kinds of PDEs and it was verified by comparing with several existing methods that it offers a very stable, highly accurate and efficient approach, see [18–20]. In 2018, Boonklurb et al. [21] modified the original FIM via Chebyshev polynomials for solving linear PDEs which provided a much higher accuracy than the FDM and those traditional FIMs. Unfortunately, the modified FIM in [21] has never been studied for the Burgers' equations and coupled Burgers' equations involving fractional order derivatives with respect to time. This became the major motivation to carry out the current work.

In this paper, we improve the modified FIM in [21] by using the shifted Chebyshev polynomials (FIM-SCP) to devise the numerical algorithms for finding the decent approximate solutions of time-fractional Burgers' equations both in one- and two-dimensional domains as well as time-fractional coupled Burgers' equations. Their time-fractional derivative terms are described in the Caputo sense. We note here that the FIM in [21] is applicable for solving linear differential equations. With our improvement in this paper, we propose the numerical methods that are applicable for solving time-fractional Burgers' equations. It is well known that Chebyshev polynomial have the orthogonal property which plays an important role in the theory of approximation. The roots of the Chebyshev polynomial can be found explicitly and when the equidistant nodes are so bad, we can overcome the problem by using the Chebyshev nodes. If we sample our function at the Chebyshev nodes, we can have best approximation under the maximum norm, see [22] for more details. With these advantages, our improved FIM-SCP is constructed by approximating the solutions expressed in term of the shifted Chebyshev expansion. We use the zeros of the Chebyshev polynomial of a certain degree to interpolate the approximate solution. With our work, we obtain the shifted Chebyshev integration matrices in one- and two- dimensional spaces which are used to deal with the spatial discretizations. The temporal discretizations are approximated by the forward difference quotient.

The rest of this paper is organized as follows. In Section 2, we provide the basic definitions and the necessary notations used throughout this paper. In Section 3, the improved FIM-SCP of constructing the shifted Chebyshev integration matrices, both for one and two dimensions are discussed. In Section 4, we derive the numerical algorithms for solving one-dimensional time-fractional Burgers' equations, two-dimensional time-fractional Burgers' equations, and time-fractional coupled Burgers' equations. The numerical results are presented, which are also shown to be more computationally efficient and accurate than the other methods with CPU time(s) and rate of convergence. The conclusion and some discussion for the future work are provided in Section 5.

2. Preliminaries

Before embarking into the details of the FIM-SCP for solving time-fractional differential equations, we provide in this section the basic definitions of fractional derivatives and shifted Chebyshev polynomials. The necessary notations and some important facts used throughout this paper are also given. More details on basic results of fractional calculus can be found in [23] and further details of Chebyshev polynomials can be reached in [22].

Definition 1. *Let p, μ, and t be real numbers such that $t > 0$, and*

$$C_\mu = \{u(t) \mid u(t) = t^p u_1(t), \text{ where } u_1(t) \in C[0,\infty) \text{ and } p > \mu\}.$$

If an integrable function $u(t) \in C_\mu$, we define the Riemann–Liouville fractional integral operator of order $\alpha \geq 0$ as

$$I^\alpha u(t) = \begin{cases} \frac{1}{\Gamma(\alpha)} \int_0^t \frac{u(s)}{(t-s)^{1-\alpha}} ds & \text{for } \alpha > 0, \\ u(t) & \text{for } \alpha = 0, \end{cases}$$

where $\Gamma(\cdot)$ is the well-known Gamma function.

Definition 2. *The Caputo fractional derivative D^α of $u(t) \in C_{-1}^m$, with $u(t) \in C_\mu^m$ if and only if $u^{(m)} \in C_\mu$, is defined by*

$$D^\alpha u(t) = I^{m-\alpha} D^m u(t) = \begin{cases} \frac{1}{\Gamma(m-\alpha)} \int_0^t \frac{u^{(m)}(s)}{(t-s)^{1-m+\alpha}} ds & \text{for } \alpha \in (m-1, m), \\ u^{(m)}(t) & \text{for } \alpha = m, \end{cases}$$

where $m \in \mathbb{N}$ and $t > 0$.

Definition 3. *The shifted Chebyshev polynomial of degree $n \geq 0$ for $L \in \mathbb{R}^+$ is defined by*

$$T_n^*(x) = \cos\left(n \arccos\left(\frac{2x}{L} - 1\right)\right) \text{ for } x \in [0, L]. \tag{1}$$

Lemma 1. *(i) For $n \in \mathbb{N}$, the zeros of the shifted Chebyshev polynomial $T_n^*(x)$ are*

$$x_k = \frac{L}{2}\left[\cos\left(\frac{2k-1}{2n}\pi\right) + 1\right], \ k \in \{1,2,3,...,n\}. \tag{2}$$

(ii) For $x \in [0, L]$, the single layer integrations of the shifted Chebyshev polynomial $T_n^*(x)$ are

$$\overline{T}_0^*(x) = \int_0^x T_0^*(\xi)\, d\xi = x,$$

$$\overline{T}_1^*(x) = \int_0^x T_1^*(\xi)\, d\xi = \frac{x^2}{L} - x,$$

$$\overline{T}_n^*(x) = \int_0^x T_n^*(\xi)\, d\xi = \frac{L}{4}\left[\frac{T_{n+1}^*(x)}{n+1} - \frac{T_{n-1}^*(x)}{n-1} - \frac{2(-1)^n}{n^2-1}\right], \ n \in \{2,3,4,...\}.$$

(iii) Let $\{x_k\}_{k=1}^n$ be a set of zeros of $T_n^*(x)$ defined in (2), and define the shifted Chebyshev matrix $\mathbf{T}$ by

$$\mathbf{T} = \begin{bmatrix} T_0^*(x_1) & T_1^*(x_1) & \cdots & T_{n-1}^*(x_1) \\ T_0^*(x_2) & T_1^*(x_2) & \cdots & T_{n-1}^*(x_2) \\ \vdots & \vdots & \ddots & \vdots \\ T_0^*(x_n) & T_1^*(x_n) & \cdots & T_{n-1}^*(x_n) \end{bmatrix}.$$

Then, it has the multiplicative inverse $\mathbf{T}^{-1} = \frac{1}{n}\text{diag}(1,2,2,...,2)\mathbf{T}^\top$.

3. Improved FIM-SCP

In this section, we improve the technique of Boonklurb et al. [21] to construct the first and higher order integration matrices in one and two dimensions. We note here that Boonklurb et al. used Chebyshev polynomials to construct the integration matrices and obtained numerical algorithms for solving linear differential equations, whereas in this work, we use the shifted Chebyshev polynomials to construct first and higher order shifted Chebyshev integration matrices to obtain numerical algorithms that are applicable to solve time-fractional Burgers' equations on any domain $[0, L]$ rather than $[-1, 1]$.

3.1. One-Dimensional Shifted Chebyshev Integration Matrices

Let $M \in \mathbb{N}$ and $L \in \mathbb{R}^+$. Define an approximate solution $u(x)$ of a certain PDE by the linear combination of shifted Chebyshev polynomials (1), i.e.,

$$u(x) = \sum_{n=0}^{M-1} c_n T_n^*(x) \text{ for } x \in [0, L]. \tag{3}$$

Let x_k, $k \in \{1,2,3,...,M\}$, be the grid points generated by the zeros of the shifted Chebyshev polynomial $T_M^*(x)$ defined in (2). Substituting each x_k into (3), then (3) can be expressed as

$$\begin{bmatrix} u(x_1) \\ u(x_2) \\ \vdots \\ u(x_M) \end{bmatrix} = \begin{bmatrix} T_0^*(x_1) & T_1^*(x_1) & \cdots & T_{M-1}^*(x_1) \\ T_0^*(x_2) & T_1^*(x_2) & \cdots & T_{M-1}^*(x_2) \\ \vdots & \vdots & \ddots & \vdots \\ T_0^*(x_M) & T_1^*(x_M) & \cdots & T_{M-1}^*(x_M) \end{bmatrix} \begin{bmatrix} c_0 \\ c_1 \\ \vdots \\ c_{M-1} \end{bmatrix},$$

and we let it be denoted by $\mathbf{u} = \mathbf{Tc}$. The coefficients $\{c_n\}_{n=0}^{M-1}$ can be obtained by computing $\mathbf{c} = \mathbf{T}^{-1}\mathbf{u}$. Let $U^{(1)}(x_k)$ denote the single layer integration of u from 0 to x_k. Then,

$$U^{(1)}(x_k) = \int_0^{x_k} u(\xi)\,d\xi = \sum_{n=0}^{M-1} c_n \int_0^{x_k} T_n^*(\xi)\,d\xi = \sum_{n=0}^{M-1} c_n \overline{T}_n^*(x_k)$$

for $k \in \{1,2,3,...,M\}$ or in matrix form:

$$\begin{bmatrix} U^{(1)}(x_1) \\ U^{(1)}(x_2) \\ \vdots \\ U^{(1)}(x_M) \end{bmatrix} = \begin{bmatrix} \overline{T}_0^*(x_1) & \overline{T}_1^*(x_1) & \cdots & \overline{T}_{M-1}^*(x_1) \\ \overline{T}_0^*(x_2) & \overline{T}_1^*(x_2) & \cdots & \overline{T}_{M-1}^*(x_2) \\ \vdots & \vdots & \ddots & \vdots \\ \overline{T}_0^*(x_M) & \overline{T}_1^*(x_M) & \cdots & \overline{T}_{M-1}^*(x_M) \end{bmatrix} \begin{bmatrix} c_0 \\ c_1 \\ \vdots \\ c_{M-1} \end{bmatrix}.$$

We denote the above equation by $\mathbf{U}^{(1)} = \overline{\mathbf{T}}\mathbf{c} = \overline{\mathbf{T}}\mathbf{T}^{-1}\mathbf{u} := \mathbf{A}\mathbf{u}$, where $\mathbf{A} = \overline{\mathbf{T}}\mathbf{T}^{-1} := [a_{ki}]_{M \times M}$ is called the "shifted Chebyshev integration matrix" for the improved FIM-SCP in one dimension. Next, let us consider the double layer integration of u from 0 to x_k that denoted by $U^{(2)}(x_k)$. We have

$$U^{(2)}(x_k) = \int_0^{x_k}\int_0^{\xi_2} u(\xi_1)\,d\xi_1 d\xi_2 = \sum_{i=1}^{M} a_{ki}\int_0^{x_i} u(\xi_1)\,d\xi_1 = \sum_{i=1}^{M}\sum_{j=1}^{M} a_{ki}a_{ij}u(x_j)$$

for $k \in \{1,2,3,...,M\}$, it can be written in matrix form as $\mathbf{U}^{(2)} = \mathbf{A}^2\mathbf{u}$. The mth layer integration of u from 0 to x_k, denoted by $U^{(m)}(x_k)$, can be obtained in the similar manner, that is,

$$U^{(m)}(x_k) = \int_0^{x_k}\cdots\int_0^{\xi_2} u(\xi_1)\,d\xi_1 \cdots d\xi_m = \sum_{i_m=1}^{M}\cdots\sum_{j=1}^{M} a_{ki_m}\cdots a_{i_1 j}u(x_j)$$

for $k \in \{1, 2, 3, ..., M\}$, or written in the matrix form as $\mathbf{U}^{(m)} = \mathbf{A}^m \mathbf{u}$.

3.2. Two-Dimensional Shifted Chebyshev Integration Matrices

Let $M, N \in \mathbb{N}$ and $L_1, L_2 \in \mathbb{R}^+$. Divide the domain $[0, L_1] \times [0, L_2]$ into a mesh with M nodes by N nodes along the horizontal and the vertical directions, respectively. Let x_k, where $k \in \{1, 2, 3, ..., M\}$, be the grid points generated by the shifted Chebyshev nodes of $T_M^*(x)$ and let y_s, where $s \in \{1, 2, 3, ..., N\}$, be the grid points generated by the shifted Chebyshev nodes of $T_N^*(y)$. Thus, there are $M \times N$ grid points in total. For computation, we index the numbering of grid points along the x-direction by the global numbering system (Figure 1a) and along y-direction by the local numbering system (Figure 1b).

Let $U_x^{(1)}$ and $U_y^{(1)}$ be the single layer integrations with respect to the variables x and y, respectively. For each fixed y, we have $U_x^{(1)}(x_k, y)$ in the global numbering system as

$$U_x^{(1)}(x_k, y) = \int_0^{x_k} u(\xi, y)\, d\xi = \sum_{i=1}^{M} a_{ki} u(x_i, y). \tag{4}$$

For $k \in \{1, 2, 3, ..., M\}$, (4) can be expressed as $\mathbf{U}_x^{(1)}(\cdot, y) = \mathbf{A}_M \mathbf{u}(\cdot, y)$, where $\mathbf{A}_M = \overline{\mathbf{T}}\mathbf{T}^{-1}$ is the $M \times M$ matrix. Thus, for each $y \in \{y_1, y_2, y_3, ..., y_N\}$,

$$\begin{bmatrix} \mathbf{U}_x^{(1)}(\cdot, y_1) \\ \mathbf{U}_x^{(1)}(\cdot, y_2) \\ \vdots \\ \mathbf{U}_x^{(1)}(\cdot, y_N) \end{bmatrix} = \underbrace{\begin{bmatrix} \mathbf{A}_M & 0 & \cdots & 0 \\ 0 & \mathbf{A}_M & \ddots & \vdots \\ \vdots & \ddots & \ddots & 0 \\ 0 & \cdots & 0 & \mathbf{A}_M \end{bmatrix}}_{N \text{ blocks}} \begin{bmatrix} \mathbf{u}(\cdot, y_1) \\ \mathbf{u}(\cdot, y_2) \\ \vdots \\ \mathbf{u}(\cdot, y_N) \end{bmatrix},$$

we shall denote it by $\mathbf{U}_x^{(1)} = \mathbf{A}_x \mathbf{u}$, where $\mathbf{A}_x = \mathbf{I}_N \otimes \mathbf{A}_M$ is the shifted Chebyshev integration matrix with respect to x-axis and $\otimes$ is the Kronecker product defined in [24]. Similarly, for each fixed x, $U_y^{(1)}(x, y_s)$ can be expressed in the local numbering system as

$$U_y^{(1)}(x, y_s) = \int_0^{y_s} u(x, \eta)\, d\eta = \sum_{j=1}^{N} a_{sj} u(x, y_j). \tag{5}$$

For $s \in \{1, 2, 3, ..., N\}$, (5) can be written as $\mathbf{U}_y^{(1)}(x, \cdot) = \mathbf{A}_N \mathbf{u}(x, \cdot)$, where $\mathbf{A}_N = \overline{\mathbf{T}}\mathbf{T}^{-1}$ is the $N \times N$ matrix. Therefore, for each $x \in \{x_1, x_2, x_3, ..., x_M\}$,

$$\begin{bmatrix} \mathbf{U}_y^{(1)}(x_1, \cdot) \\ \mathbf{U}_y^{(1)}(x_2, \cdot) \\ \vdots \\ \mathbf{U}_y^{(1)}(x_M, \cdot) \end{bmatrix} = \underbrace{\begin{bmatrix} \mathbf{A}_N & 0 & \cdots & 0 \\ 0 & \mathbf{A}_N & \ddots & \vdots \\ \vdots & \ddots & \ddots & 0 \\ 0 & \cdots & 0 & \mathbf{A}_N \end{bmatrix}}_{M \text{ blocks}} \begin{bmatrix} \mathbf{u}(x_1, \cdot) \\ \mathbf{u}(x_2, \cdot) \\ \vdots \\ \mathbf{u}(x_M, \cdot) \end{bmatrix}.$$

We shall denote the above matrix equation by $\widetilde{\mathbf{U}}_y^{(1)} = \widetilde{\mathbf{A}}_y \widetilde{\mathbf{u}}$, where $\widetilde{\mathbf{A}}_y = \mathbf{I}_M \otimes \mathbf{A}_N$. We notice that the elements of $\mathbf{u}$ and $\widetilde{\mathbf{u}}$ are the same but different positions in the numbering system. Thus, we can transform $\tilde{\mathbf{U}}_y^{(1)}$ and $\tilde{\mathbf{u}}$ in the local numbering system to the global numbering system by using the permutation matrix $\mathbf{P} = [p_{ij}]_{MN \times MN}$, where each p_{ij} is defined by

$$p_{ij} = \begin{cases} 1 & ; \begin{cases} i = (s-1)M + k, \\ j = (k-1)N + s, \end{cases} \\ 0 & ; \text{otherwise,} \end{cases} \qquad (6)$$

for all $k \in \{1,2,3,...,M\}$ and $s \in \{1,2,3,...,N\}$. We obtain that $\mathbf{U}_y^{(1)} = \mathbf{P}\tilde{\mathbf{U}}_y^{(1)}$ and $\mathbf{u} = \mathbf{P}\tilde{\mathbf{u}}$. Therefore, we have $\mathbf{U}_y^{(1)} = \mathbf{A}_y \mathbf{u}$, where $\mathbf{A}_y = \mathbf{P}\tilde{\mathbf{A}}_y \mathbf{P}^{-1} = \mathbf{P}(\mathbf{I}_M \otimes \mathbf{A}_N)\mathbf{P}^\top$ is the shifted Chebyshev integration matrix with respect to y-axis in the global numbering system.

Remark 1 ([21]). *Let $m, n \in \mathbb{N}$, the multi-layer integrations in the global numbering system can be represented in the matrix forms as follows,*

- *the mth layer integration with respect to x is $\mathbf{U}_x^{(m)} = \mathbf{A}_x^m \mathbf{u}$,*
- *the nth layer integration with respect to y is $\mathbf{U}_y^{(n)} = \mathbf{A}_y^n \mathbf{u}$,*
- *the multi-layer integration with respect to both x and y is $\mathbf{U}_{xy}^{(m,n)} = \mathbf{A}_x^m \mathbf{A}_y^n \mathbf{u}$.*

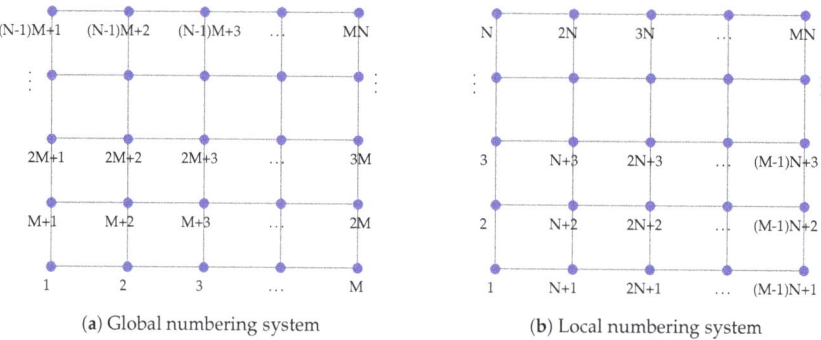

Figure 1. Global and local grid points.

4. The Numerical Algorithms for Time-Fractional Burgers' Equations

In this section, we derive the numerical algorithms based on our improved FIM-SCP for solving time-fractional Burgers' equations both in one and two dimensions. The numerical algorithm for solving time-fractional coupled Burgers' equations is also proposed. To demonstrate the effectiveness and the efficiency of our algorithms, some numerical examples are given. Moreover, we find the time convergence rates and CPU times(s) of each example in order to demonstrate the computational cost. We note here that we implemented our numerical algorithms in MatLab R2016a. The experimental computer system is configured as: Intel(R) Core(TM) i7-6700 CPU @ 3.40 GHz. Finally, the graphically numerical solutions of each example are also depicted.

4.1. Algorithm for One-Dimensional Time-Fractional Burgers' Equation

Let L and T be positive real numbers and $\alpha \in (0,1]$. Consider the time-fractional Burgers' equation with a viscosity parameter $\nu > 0$ as follows.

$$\frac{\partial^\alpha u}{\partial t^\alpha} + u \frac{\partial u}{\partial x} - \nu \frac{\partial^2 u}{\partial x^2} = f(x,t), \quad x \in (0,L), \ t \in (0,T], \qquad (7)$$

subject to the initial condition
$$u(x,0) = \phi(x), \quad x \in [0, L], \tag{8}$$

and the boundary conditions
$$u(0,t) = \psi_1(t) \text{ and } u(L,t) = \psi_2(t), \quad t \in (0, T], \tag{9}$$

where $f(x,t)$, $\phi(x)$, $\psi_1(t)$, and $\psi_2(t)$ are given functions. Let us first linearize (7) by determining the iteration at time $t_m = m(\Delta t)$, where Δt is the time step and $m \in \mathbb{N}$. Then, we have

$$\left.\frac{\partial^\alpha u}{\partial t^\alpha}\right|_{t=t_m} + u^{m-1}\frac{\partial u^m}{\partial x} - \nu\frac{\partial^2 u^m}{\partial x^2} = f(x, t_m), \tag{10}$$

where $u^m = u(x, t_m)$ is the numerical solution at the mth iteration. For the Caputo time-fractional derivative term defined in Definition 2, we have

$$\left.\frac{\partial^\alpha u}{\partial t^\alpha}\right|_{t=t_m} = \frac{1}{\Gamma(1-\alpha)} \int_0^{t_m} \frac{u_s(x,s)}{(t_m-s)^\alpha} ds = \frac{1}{\Gamma(1-\alpha)} \sum_{i=0}^{m-1} \int_{t_i}^{t_{i+1}} \frac{u_s(x,s)}{(t_m-s)^\alpha} ds. \tag{11}$$

Using the first-order forward difference quotient to approximate the derivative term in (11), we get

$$\left.\frac{\partial^\alpha u}{\partial t^\alpha}\right|_{t=t_m} \approx \frac{1}{\Gamma(1-\alpha)} \sum_{i=0}^{m-1} \int_{t_i}^{t_{i+1}} (t_m-s)^{-\alpha} \left(\frac{u^{i+1}-u^i}{\Delta t}\right) ds$$

$$= \frac{1}{\Gamma(1-\alpha)} \sum_{i=0}^{m-1} \left(\frac{u^{i+1}-u^i}{\Delta t}\right) \left[\frac{(t_m-t_i)^{1-\alpha} - (t_m-t_{i+1})^{1-\alpha}}{1-\alpha}\right]$$

$$= \frac{1}{\Gamma(2-\alpha)} \sum_{i=0}^{m-1} \left(\frac{u^{i+1}-u^i}{\Delta t}\right) \left[(m-i)^{1-\alpha} - (m-i-1)^{1-\alpha}\right] (\Delta t)^{1-\alpha}$$

$$= \frac{(\Delta t)^{-\alpha}}{\Gamma(2-\alpha)} \sum_{j=0}^{m-1} (u^{m-j} - u^{m-j-1}) \left[(j+1)^{1-\alpha} - j^{1-\alpha}\right]$$

$$= \sum_{j=0}^{m-1} w_j (u^{m-j} - u^{m-j-1}), \tag{12}$$

where $w_j = \frac{(\Delta t)^{-\alpha}}{\Gamma(2-\alpha)}\left[(j+1)^{1-\alpha} - j^{1-\alpha}\right]$. Thus, (10) becomes

$$w_0(u^m - u^{m-1}) + \sum_{j=1}^{m-1} w_j(u^{m-j} - u^{m-j-1}) + u^{m-1}\frac{\partial u^m}{\partial x} - \nu\frac{\partial^2 u^m}{\partial x^2} = f(x, t_m). \tag{13}$$

In order to eliminate the derivative terms in (13), we apply the modified FIM by taking the double layer integration. Then, for each shifted Chebyshev node x_k, $k \in \{1,2,3,...,M\}$, we obtain

$$w_0 \int_0^{x_k}\int_0^\eta (u^m - u^{m-1}) d\xi d\eta + \sum_{j=1}^{m-1} w_j \int_0^{x_k}\int_0^\eta (u^{m-j} - u^{m-j-1}) d\xi d\eta$$

$$+ \int_0^{x_k}\int_0^\eta \left(u^{m-1}\frac{\partial u^m}{\partial \xi}\right) d\xi d\eta - \nu u^m + d_1 x_k + d_2 = \int_0^{x_k}\int_0^\eta f(\xi, t_m) d\xi d\eta, \tag{14}$$

where d_1 and d_2 are the arbitrary constants of integration. Next, we consider the nonlinear term in (14). By using the technique of integration by parts, we have

$$q(x_k) := \int_0^{x_k} \int_0^\eta \left(u^{m-1}\frac{\partial u^m}{\partial \xi}\right) d\xi d\eta$$

$$= \int_0^{x_k} u^{m-1}(\eta)u^m(\eta)\,d\eta - \int_0^{x_k}\int_0^\eta \frac{\partial u^{m-1}(\xi)}{\partial \xi} u^m(\xi)\,d\xi d\eta$$

$$= \int_0^{x_k} u^{m-1}(\eta)u^m(\eta)\,d\eta - \int_0^{x_k}\int_0^\eta \sum_{n=0}^{M-1} c_n^{m-1} \frac{dT_n^*(\xi)}{d\xi} u^m(\xi)\,d\xi d\eta$$

$$= \int_0^{x_k} u^{m-1}(\eta)u^m(\eta)\,d\eta - \int_0^{x_k}\int_0^\eta \mathbf{T}'(\xi)\mathbf{T}^{-1}\mathbf{u}^{m-1} u^m(\xi)\,d\xi d\eta, \qquad (15)$$

where $\mathbf{T}'(\xi) = \left[\frac{dT_0^*(\xi)}{d\xi}, \frac{dT_1^*(\xi)}{d\xi}, \frac{dT_2^*(\xi)}{d\xi}, ..., \frac{dT_{M-1}^*(\xi)}{d\xi}\right]$. Thus, for $k \in \{1,2,3,...,M\}$, (15) can be expressed in matrix form as

$$\begin{bmatrix} q(x_1) \\ q(x_2) \\ \vdots \\ q(x_M) \end{bmatrix} = \mathbf{A} \begin{bmatrix} u^{m-1}(x_1)u^m(x_1) \\ u^{m-1}(x_2)u^m(x_2) \\ \vdots \\ u^{m-1}(x_M)u^m(x_M) \end{bmatrix} - \mathbf{A}^2 \begin{bmatrix} \mathbf{T}'(x_1)\mathbf{T}^{-1}\mathbf{u}^{m-1}u^m(x_1) \\ \mathbf{T}'(x_2)\mathbf{T}^{-1}\mathbf{u}^{m-1}u^m(x_2) \\ \vdots \\ \mathbf{T}'(x_M)\mathbf{T}^{-1}\mathbf{u}^{m-1}u^m(x_M) \end{bmatrix}.$$

For computational convenience, we reduce the above equation into the matrix form:

$$\mathbf{q} = \mathbf{A}\,\mathrm{diag}\left(\mathbf{u}^{m-1}\right)\mathbf{u}^m - \mathbf{A}^2\,\mathrm{diag}\left(\mathbf{T}'\mathbf{T}^{-1}\mathbf{u}^{m-1}\right)\mathbf{u}^m := \mathbf{Q}\mathbf{u}^m, \qquad (16)$$

where $\mathbf{q} = [q(x_1), q(x_2), q(x_3)..., q(x_M)]$, $\mathbf{Q} = \mathbf{A}\,\mathrm{diag}(\mathbf{u}^{m-1}) - \mathbf{A}^2\,\mathrm{diag}(\mathbf{T}'\mathbf{T}^{-1}\mathbf{u}^{m-1})$, and

$$\mathbf{T}' = \begin{bmatrix} \mathbf{T}'(x_1) \\ \mathbf{T}'(x_2) \\ \vdots \\ \mathbf{T}'(x_M) \end{bmatrix} = \begin{bmatrix} \frac{dT_0^*(\xi)}{d\xi}\big|_{x_1} & \frac{dT_1^*(\xi)}{d\xi}\big|_{x_1} & \cdots & \frac{dT_{M-1}^*(\xi)}{d\xi}\big|_{x_1} \\ \frac{dT_0^*(\xi)}{d\xi}\big|_{x_2} & \frac{dT_1^*(\xi)}{d\xi}\big|_{x_2} & \cdots & \frac{dT_{M-1}^*(\xi)}{d\xi}\big|_{x_2} \\ \vdots & \vdots & \ddots & \vdots \\ \frac{dT_0^*(\xi)}{d\xi}\big|_{x_M} & \frac{dT_1^*(\xi)}{d\xi}\big|_{x_M} & \cdots & \frac{dT_{M-1}^*(\xi)}{d\xi}\big|_{x_M} \end{bmatrix}. \qquad (17)$$

Consequently, for $k \in \{1,2,3,...,M\}$ by hiring (16) and the idea of Boonklurb et al. [21], we can convert (14) into the matrix form as

$$w_0\mathbf{A}^2(\mathbf{u}^m - \mathbf{u}^{m-1}) + \sum_{j=1}^{m-1} w_j\mathbf{A}^2(\mathbf{u}^{m-j} - \mathbf{u}^{m-j-1}) + \mathbf{Q}\mathbf{u}^m - \nu\mathbf{u}^m + d_1\mathbf{x} + d_2\mathbf{i} = \mathbf{A}^2\mathbf{f}^m$$

$$\left[w_0\mathbf{A}^2 + \mathbf{Q} - \nu\mathbf{I}\right]\mathbf{u}^m + d_1\mathbf{x} + d_2\mathbf{i} = \mathbf{A}^2\mathbf{f}^m + w_0\mathbf{A}^2\mathbf{u}^{m-1} - \sum_{j=1}^{m-1} w_j\mathbf{A}^2(\mathbf{u}^{m-j} - \mathbf{u}^{m-j-1}), \qquad (18)$$

where $\mathbf{I}$ is the $M \times M$ identity matrix, $\mathbf{i} = [1,1,1,...,1]^\top$, $\mathbf{u}^m = [u(x_1,t_m), u(x_2,t_m), ..., u(x_M,t_m)]^\top$, $\mathbf{x} = [x_1, x_2, x_3, ..., x_M]^\top$, $\mathbf{f}^m = [f(x_1,t_m), f(x_2,t_m), ..., f(x_M,t_m)]^\top$ and $\mathbf{A} = \overline{\mathbf{T}}\mathbf{T}^{-1}$. For the boundary conditions (9), we can change them into the vector forms by using the linear combination of the shifted Chebyshev polynomial at the mth iteration as follows.

$$u(0,t_m) = \sum_{n=0}^{M-1} c_n^m T_n^*(0) = \sum_{n=0}^{M-1} c_n^m(-1)^n := \mathbf{t}_l\mathbf{c}^m = \mathbf{t}_l\mathbf{T}^{-1}\mathbf{u}^m = \psi_1(t_m), \qquad (19)$$

$$u(L,t_m) = \sum_{n=0}^{M-1} c_n^m T_n^*(L) = \sum_{n=0}^{M-1} c_n^m(1)^n := \mathbf{t}_r\mathbf{c}^m = \mathbf{t}_r\mathbf{T}^{-1}\mathbf{u}^m = \psi_2(t_m), \qquad (20)$$

where $\mathbf{t}_l = [1, -1, 1, ..., (-1)^{M-1}]$ and $\mathbf{t}_r = [1, 1, 1, ..., 1]$.

From (18)–(20), we can construct the following system of iterative linear equations that contains $M+2$ unknowns

$$\begin{bmatrix} w_0\mathbf{A}^2 + \mathbf{Q} - \nu\mathbf{I} & \mathbf{x} & \mathbf{i} \\ \mathbf{t}_l\mathbf{T}^{-1} & 0 & 0 \\ \mathbf{t}_r\mathbf{T}^{-1} & 0 & 0 \end{bmatrix} \begin{bmatrix} \mathbf{u}^m \\ d_1 \\ d_2 \end{bmatrix} = \begin{bmatrix} \mathbf{A}^2\mathbf{f}^m + w_0\mathbf{A}^2\mathbf{u}^{m-1} - \mathbf{s} \\ \psi_1(t_m) \\ \psi_2(t_m) \end{bmatrix}, \qquad (21)$$

where $\mathbf{s} = \sum_{j=1}^{m-1} w_j \mathbf{A}^2 (\mathbf{u}^{m-j} - \mathbf{u}^{m-j-1})$ for $m > 1$, and $\mathbf{s} = \mathbf{0}$ if $m = 1$. Thus, starting from the initial condition $\mathbf{u}^0 = [\phi(x_1), \phi(x_2), \phi(x_3), ..., \phi(x_M)]^\top$, the approximate solution $\mathbf{u}^m$ can be obtained by solving the system (21). We note here that, for any fixed $t \in (0, T]$, the approximate solution $u(x,t)$ for each arbitrary $x \in [0, L]$ can be computed from

$$u(x,t) = \sum_{n=0}^{M-1} c_n T_n^*(x) = \mathbf{t}_x \mathbf{c}^m = \mathbf{t}_x \mathbf{T}^{-1} \mathbf{u}^m,$$

where $\mathbf{t}_x = [T_0^*(x), T_1^*(x), T_2^*(x), ..., T_{M-1}^*(x)]$ and $\mathbf{u}^m$ is the final iterative solution of (21).

Example 1. *Consider the time-fractional Burgers' Equation (7) for $x \in (0,1)$ and $t \in (0,1]$ with*

$$f(x,t) = \frac{2t^{2-\alpha}e^x}{\Gamma(3-\alpha)} + t^4 e^{2x} - \nu t^2 e^x,$$

subject to the initial condition

$$u(x,0) = 0, \; x \in [0,1]$$

and the boundary conditions

$$u(0,t) = t^2, \; u(1,t) = et^2, \; t \in (0,1].$$

The exact solution given by Esen and Tasbozan [12] is $u^*(x,t) = t^2 e^x$. In the numerical test, we choose the kinematic viscosity $\nu = 1$, $\alpha = 0.5$ and $\Delta t = 0.00025$. Table 1 presents the exact solution $u^*(x,1)$, the numerical solution $u(x,1)$ by using our FIM-SCP in Algorithm 1, and the solution obtained by the quadratic B-spline finite element Galerkin method (QBS-FEM) proposed by Esen and Tasbozan [12]. The comparison between the absolute errors E_a (as the difference in absolute value between the approximate solution and the exact solution) of the two methods shows that our FIM-SCP is more accurate than QBS-FEM for $M = 10$ and similar accuracy for other M. Algorithm 1 acquires the significant improvement in accuracy with less computational nodal points M and regardless the time steps Δt and the fractional order derivatives α. With the selection of $\alpha = 0.5$ and $M = 40$, Table 2 shows the comparison between the exact solution $u^*(x,1)$ and the numerical solution $u(x,1)$ using Algorithm 1 for various values of $\Delta t \in \{0.05, 0.01, 0.005, 0.001\}$. Table 3 illustrates the comparison between the exact solution $u^*(x,1)$ and the numerical solution $u(x,1)$ by our method for $\Delta t = 0.001$, $M = 40$, and $\alpha \in \{0.1, 0.25, 0.75, 0.9\}$. Moreover, the convergence rates are estimated by using our FIM-SCP with the discretization points $M = 20$ and step sizes $\Delta t = 2^{-k}$ for $k \in \{4, 5, 6, 7, 8\}$. In Table 4, we observe that these time convergence rates for the ℓ^∞ norm indeed are almost $O(\Delta t)$ for the different $\alpha \in (0,1)$. Then, we also find the computational cost in term of CPU time(s) in Table 4. Finally, the graph of our approximate solutions, $u(x,t)$, for different times, t, and the surface plot of the solution under the parameters $\nu = 1$, $M = 40$, and $\Delta t = 0.001$, are provided in Figure 2.

Algorithm 1 The numerical algorithm for solving one-dimensional time-fractional Burgers' equation

Input: $\alpha, \nu, x, L, T, M, \Delta t, \phi(x), \psi_1(t), \psi_2(t)$, and $f(x,t)$.
Output: An approximate solution $u(x,T)$.
1: Set $x_k = \frac{L}{2}\left[\cos\left(\frac{2k-1}{2M}\pi\right) + 1\right]$ for $k \in \{1, 2, 3, ..., M\}$.
2: Compute $\mathbf{x}, \mathbf{i}, \mathbf{A}, \mathbf{t}_l, \mathbf{t}_r, \mathbf{t}_x, \mathbf{I}, \mathbf{T}, \overline{\mathbf{T}}, \mathbf{T}^{-1}$ and $\mathbf{u}^0$.
3: Set $t_0 = 0$ and $m = 0$.
4: **while** $t_m \leq T$ **do**
5:     Set $m = m + 1$.
6:     Set $t_m = m\Delta t$.
7:     Set $\mathbf{s} = 0$.
8:     **for** $j = 1$ to $m - 1$ **do**
9:         Compute $w_j = \frac{(\Delta t)^{-\alpha}}{\Gamma(2-\alpha)}\left[(j+1)^{1-\alpha} - j^{1-\alpha}\right]$.
10:        Compute $\mathbf{s} = \mathbf{s} + w_j \mathbf{A}^2(\mathbf{u}^{m-j} - \mathbf{u}^{m-j-1})$.
11:     **end for**
12:     Compute $\mathbf{f}^m = [f(x_1, t_m), f(x_2, t_m), f(x_3, t_m), ..., f(x_M, t_m)]^\top$.
13:     Find $\mathbf{u}^m$ by solving the iterative linear system (21).
14: **end while**
15: **return** $u(x, T) = \mathbf{t}_x \mathbf{T}^{-1} \mathbf{u}^m$.

Table 1. Comparison of absolute errors E_a between QBS-FEM and FIM-SCP for Example 1.

M	x	$u^*(x,1)$	QBS-FEM [12]		FIM-SCP Algorithm 1	
			$u(x,1)$	E_a	$u(x,1)$	E_a
10	0.2	1.221403	1.222203	8.00×10^{-4}	1.221462	5.9578×10^{-5}
	0.4	1.491825	1.493437	1.61×10^{-3}	1.491934	1.0910×10^{-4}
	0.6	1.822119	1.824294	2.18×10^{-3}	1.822258	1.3933×10^{-4}
	0.8	2.225541	2.227650	2.11×10^{-3}	2.225666	1.2511×10^{-4}
20	0.2	1.221403	1.221644	2.41×10^{-4}	1.221462	5.9578×10^{-5}
	0.4	1.491825	1.492287	4.62×10^{-4}	1.491934	1.0910×10^{-4}
	0.6	1.822119	1.822727	6.08×10^{-4}	1.822258	1.3933×10^{-4}
	0.8	2.225541	2.226118	5.77×10^{-4}	2.225666	1.2511×10^{-4}
40	0.2	1.221403	1.221493	9.00×10^{-5}	1.221462	5.9578×10^{-5}
	0.4	1.491825	1.491996	1.71×10^{-4}	1.491934	1.0910×10^{-4}
	0.6	1.822119	1.822342	2.03×10^{-4}	1.822258	1.3933×10^{-4}
	0.8	2.225541	2.225747	2.06×10^{-4}	2.225666	1.2511×10^{-4}

Table 2. Absolute errors E_a at different Δt for Example 1 by FIM-SCP with $\alpha = 0.5$ and $M = 40$.

x	$u^*(x,1)$	$\Delta t = 0.05$		$\Delta t = 0.01$		$\Delta t = 0.005$		$\Delta t = 0.001$	
		$u(x,1)$	E_a	$u(x,1)$	E_a	$u(x,1)$	E_a	$u(x,1)$	E_a
0.1	1.1051	1.1116	6.44×10^{-3}	1.1064	1.25×10^{-3}	1.1057	6.22×10^{-4}	1.1052	1.23×10^{-4}
0.3	1.3498	1.3677	1.78×10^{-2}	1.3533	3.48×10^{-3}	1.3515	1.73×10^{-3}	1.3502	3.44×10^{-4}
0.5	1.6487	1.6750	2.63×10^{-2}	1.6538	5.17×10^{-3}	1.6512	2.57×10^{-3}	1.6492	5.11×10^{-4}
0.7	2.0137	2.0423	2.86×10^{-2}	2.0194	5.67×10^{-3}	2.0165	2.82×10^{-3}	2.0143	5.63×10^{-4}
0.9	2.4596	2.4763	1.67×10^{-2}	2.4629	3.36×10^{-3}	2.4612	1.68×10^{-3}	2.4599	3.35×10^{-4}

Table 3. Absolute errors E_a at different α for Example 1 by FIM-SCP with $\Delta t = 0.001$ and $M = 40$.

x	$u^*(x,1)$	$\alpha = 0.1$		$\alpha = 0.25$		$\alpha = 0.75$		$\alpha = 0.9$	
		$u(x,1)$	E_a	$u(x,1)$	E_a	$u(x,1)$	E_a	$u(x,1)$	E_a
0.1	1.1051	1.1053	1.28×10^{-4}	1.1052	1.26×10^{-4}	1.1052	1.24×10^{-4}	1.1053	1.37×10^{-4}
0.3	1.3498	1.3502	3.60×10^{-4}	1.3502	3.54×10^{-4}	1.3502	3.45×10^{-4}	1.3502	3.77×10^{-4}
0.5	1.6487	1.6492	5.34×10^{-4}	1.6492	5.26×10^{-4}	1.6492	5.11×10^{-4}	1.6492	5.54×10^{-4}
0.7	2.0137	2.0143	5.86×10^{-4}	2.0143	5.77×10^{-4}	2.0143	5.62×10^{-4}	2.0143	6.05×10^{-4}
0.9	2.4596	2.4599	3.46×10^{-4}	2.4599	3.42×10^{-4}	2.4599	3.35×10^{-4}	2.4599	3.58×10^{-4}

Table 4. Time convergence rates and CPU time(s) for Example 1 by FIM-SCP with $M = 20$.

Δt	$\alpha = 0.1$			$\alpha = 0.5$			$\alpha = 0.9$		
	$\|\|u^* - u\|\|_\infty$	Rate	Time(s)	$\|\|u^* - u\|\|_\infty$	Rate	Time(s)	$\|\|u^* - u\|\|_\infty$	Rate	Time(s)
2^{-4}	3.65×10^{-2}	1.1926	0.2502	3.60×10^{-2}	1.1912	0.2879	4.04×10^{-2}	1.1607	0.2651
2^{-5}	1.83×10^{-2}	1.0890	0.2195	1.79×10^{-2}	1.0902	0.2014	2.00×10^{-2}	1.0770	0.1979
2^{-6}	9.18×10^{-3}	1.0438	0.4783	8.92×10^{-3}	1.0448	0.5042	9.88×10^{-3}	1.0379	0.4535
2^{-7}	4.59×10^{-3}	1.0217	1.3092	4.44×10^{-3}	1.0221	1.4068	4.88×10^{-3}	1.0189	1.3392
2^{-8}	2.30×10^{-3}	1.0108	4.3165	2.21×10^{-3}	1.0110	4.5113	2.42×10^{-3}	1.0097	4.6495

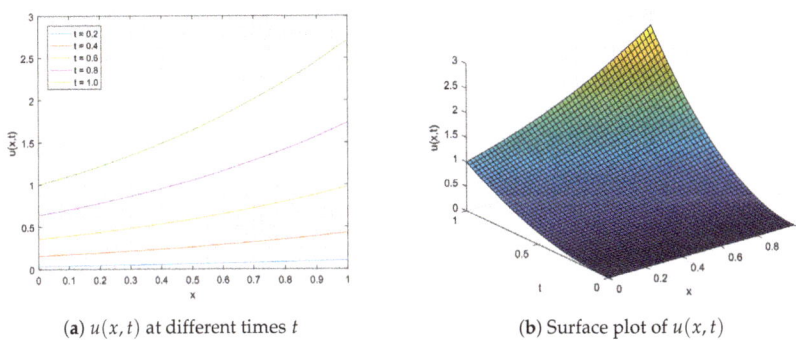

(a) $u(x,t)$ at different times t

(b) Surface plot of $u(x,t)$

Figure 2. The graphical results of Example 1 for $\nu = 1$, $M = 40$, and $\Delta t = 0.001$.

Example 2. *Consider the time-fractional Burgers' Equation (7) over $(0,1) \times (0,1]$ with $f(x,t) = 0$, subject to the initial condition*

$$u(x,0) = \left[-1 + 5\cosh\left(\frac{x}{2}\right) - 5\sinh\left(\frac{x}{2}\right)\right]^{-1}, \quad x \in [0,1],$$

and the boundary conditions

$$u(0,t) = \left[5e^{-\frac{t^\alpha}{4\Gamma(1+\alpha)}} - 1\right]^{-1} \text{ and } u(1,t) = \left[5e^{-\left(\frac{1}{2}+\frac{t^\alpha}{4\Gamma(1+\alpha)}\right)} - 1\right]^{-1}, \quad t \in (0,1].$$

The exact solution given by Yokus and Kaya [13] is $u^*(x,t) = \left[5e^{-\left(\frac{x}{2}+\frac{t^\alpha}{4\Gamma(1+\alpha)}\right)} - 1\right]^{-1}$. In our numerical test, we choose the kinematic viscosity $\nu = 1$, $\alpha = 0.8$, $M = 50$ and $\Delta t = 0.001$. Table 5 presents the exact solution $u^*(x, 0.02)$, the numerical solution $u(x, 0.02)$ by using our FIM-SCP in Algorithm 1, and the solution obtained by using the expansion method and the Cole–Hopf transformation (EPM-CHT) proposed by Yokus and Kaya in [13]. The error norms L_2 and L_∞ of this problem between our FIM-SCP and EPM-CHT with $\alpha = 0.8$ for the various values of nodal grid points $M \in \{5, 10, 20, 25, 50\}$ and step size $\Delta t = 1/M$ are illustrated in Table 6. We see that our Algorithm 1 achieves improved accuracy with less computational cost. Furthermore, we estimate the convergence rates of time for this problem by using our FIM-SCP with the discretization nodes $M = 20$ and step sizes $\Delta t = 2^{-k}$ for $k \in \{4,5,6,7,8\}$ which are tabulated in Table 7. We observe that these rates of convergence for the ℓ^∞ norm indeed are almost linear convergence $O(\Delta t)$ for the different values $\alpha \in (0,1)$. Then, we also calculate the computational cost in term of CPU time(s) as shown in Table 7. Figure 3a,b depict the numerical solutions $u(x,t)$ at different times t and the surface plot of $u(x,t)$, respectively.

Table 5. Comparison of the exact and numerical solutions for Example 2 for $\alpha = 0.8$ and $M = 50$.

x	$u^*(x, 0.02)$	EPM-CHT [13]		FIM-SCP Algorithm 1	
		$u(x, 0.02)$	E_a	$u(x, 0.02)$	E_a
0.02	0.256906	0.256321	5.84566×10^{-4}	0.256913	6.7146×10^{-6}
0.04	0.260159	0.259566	5.93809×10^{-4}	0.260173	1.3390×10^{-5}
0.06	0.263463	0.262860	6.03243×10^{-4}	0.263483	2.0005×10^{-5}
0.08	0.266817	0.266204	6.12874×10^{-4}	0.266844	2.6539×10^{-5}
0.10	0.270223	0.269601	6.22707×10^{-4}	0.270256	3.2970×10^{-5}

Table 6. Comparison of the error norms L_2 and L_∞ for Example 2 with $\alpha = 0.8$ and $\Delta t = 1/M$.

M	EPM-CHT [13]		FIM-SCP Algorithm 1		
	L_2	L_∞	L_1	L_2	L_∞
5	4.2568×10^{-2}	7.0345×10^{-2}	3.6257×10^{-4}	1.8745×10^{-4}	1.1494×10^{-5}
10	4.2708×10^{-3}	6.3200×10^{-3}	1.4701×10^{-4}	5.1150×10^{-5}	2.1754×10^{-5}
20	1.1366×10^{-3}	1.9300×10^{-3}	2.9688×10^{-4}	7.2352×10^{-5}	2.1754×10^{-5}
25	7.8890×10^{-4}	1.4410×10^{-4}	3.7153×10^{-4}	8.0893×10^{-5}	2.1754×10^{-5}
50	2.7690×10^{-4}	6.6400×10^{-4}	7.4421×10^{-4}	1.1440×10^{-5}	2.1755×10^{-5}

Table 7. Time convergence rates and CPU time(s) for Example 2 by FIM-SCP with $M = 20$.

Δt	$\alpha = 0.1$			$\alpha = 0.5$			$\alpha = 0.9$		
	$\|u^* - u\|_\infty$	Rate	Time(s)	$\|u^* - u\|_\infty$	Rate	Time(s)	$\|u^* - u\|_\infty$	Rate	Time(s)
2^{-4}	3.25×10^{-3}	1.0396	0.2123	1.22×10^{-2}	0.9895	0.2128	1.88×10^{-2}	1.0299	0.2052
2^{-5}	3.39×10^{-3}	1.0106	0.3159	6.05×10^{-3}	0.9951	0.3192	9.44×10^{-3}	1.0150	0.2836
2^{-6}	4.25×10^{-3}	1.0037	0.4858	3.01×10^{-3}	0.9976	0.4624	4.74×10^{-3}	1.0075	0.4753
2^{-7}	4.41×10^{-3}	1.0015	1.4507	2.91×10^{-3}	1.0089	1.4495	2.37×10^{-3}	1.0037	1.4213
2^{-8}	4.50×10^{-3}	1.0007	4.7479	3.35×10^{-3}	1.0037	4.3760	1.18×10^{-3}	1.0019	4.5449

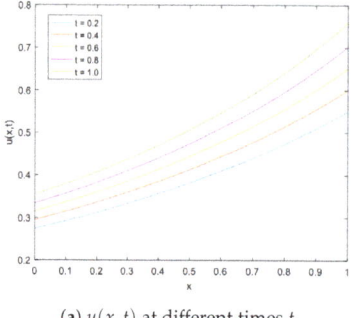

(a) $u(x,t)$ at different times t

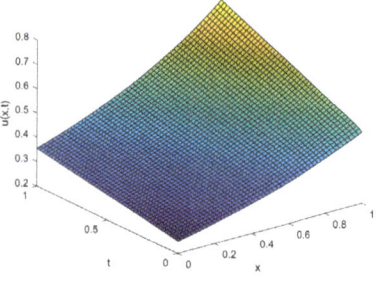

(b) Surface plot of $u(x,t)$

Figure 3. The graphical solutions of Example 2 for $\nu = 1$, $M = 40$, and $\Delta t = 0.001$.

4.2. Algorithm for Two-Dimensional Time-Fractional Burgers' Equation

Let L_1 and L_2 be positive real numbers, $\Omega = (0, L_1) \times (0, L_2)$, and $\alpha \in (0, 1]$. Consider the two-dimensional time-fractional Burgers' equation with a viscosity $\nu > 0$,

$$\frac{\partial^\alpha u}{\partial t^\alpha} + u \left(\frac{\partial u}{\partial x} + \frac{\partial u}{\partial y} \right) - \nu \left(\frac{\partial^2 u}{\partial x^2} + \frac{\partial^2 u}{\partial y^2} \right) = f(x, y, t), \quad (x, y) \in \Omega, \ t \in (0, T], \tag{22}$$

subject to the initial condition

$$u(x, y, 0) = \phi(x, y), \quad (x, y) \in \Omega, \tag{23}$$

and the boundary conditions

$$\begin{aligned}
u(0, y, t) &= \psi_1(y, t), \ u(L_1, y, t) = \psi_2(y, t), \ y \in [0, L_2], \ t \in (0, T], \\
u(x, 0, t) &= \psi_3(x, t), \ u(x, L_2, t) = \psi_4(x, t), \ x \in [0, L_1], \ t \in (0, T],
\end{aligned} \tag{24}$$

where f, ϕ, ψ_1, ψ_2, ψ_3, and ψ_4 are given functions. As $\frac{\partial}{\partial x}(\frac{u^2}{2}) = u\frac{\partial u}{\partial x}$ and $\frac{\partial}{\partial y}(\frac{u^2}{2}) = u\frac{\partial u}{\partial y}$, we can transform (22) to

$$\frac{\partial^\alpha u}{\partial t^\alpha} + \frac{\partial}{\partial x}\left(\frac{u^2}{2}\right) + \frac{\partial}{\partial y}\left(\frac{u^2}{2}\right) - \nu\left(\frac{\partial^2 u}{\partial x^2} + \frac{\partial^2 u}{\partial y^2}\right) = f(x, y, t). \tag{25}$$

Let us linearize (25) by imposing the iteration at time $t_m = m(\Delta t)$ for $m \in \mathbb{N}$ and Δt is an arbitrary time step. Thus, we have

$$\left.\frac{\partial^\alpha u}{\partial t^\alpha}\right|_{t=t_m} + \frac{\partial}{\partial x}\left(\frac{u^{m-1}}{2}u^m\right) + \frac{\partial}{\partial y}\left(\frac{u^{m-1}}{2}u^m\right) - \nu\left(\frac{\partial^2 u^m}{\partial x^2} + \frac{\partial^2 u^m}{\partial y^2}\right) = f^m, \tag{26}$$

where $f^m = f(x, y, t_m)$ and $u^m = u(x, y, t_m)$ is the numerical solution at the mth iteration. Next, consider the fractional order derivative in the Caputo sense as defined in Definition 2, by using (12), then (26) becomes

$$\sum_{j=0}^{m-1} w_j (u^{m-j} - u^{m-j-1}) + \frac{\partial}{\partial x}\left(\frac{u^{m-1}}{2}u^m\right) + \frac{\partial}{\partial y}\left(\frac{u^{m-1}}{2}u^m\right) - \nu\left(\frac{\partial^2 u^m}{\partial x^2} + \frac{\partial^2 u^m}{\partial y^2}\right) = f^m,$$

where $w_j = \frac{(\Delta t)^{-\alpha}}{\Gamma(2-\alpha)}\left[(j+1)^{1-\alpha} - j^{1-\alpha}\right]$. The above equation can be transformed to the integral equation by taking twice integrations over both x and y, we have

$$\begin{aligned}
&\sum_{j=0}^{m-1} w_j \int_0^y \int_0^{\eta_2} \int_0^x \int_0^{\xi_2} (u^{m-j} - u^{m-j-1}) d\xi_1 d\xi_2 d\eta_1 d\eta_2 + \frac{1}{2}\int_0^y \int_0^{\eta_2} \int_0^x (u^{m-1}u^m) d\xi_2 d\eta_1 d\eta_2 \\
&+ \frac{1}{2}\int_0^y \int_0^x \int_0^{\xi_2} (u^{m-1}u^m) d\xi_1 d\xi_2 d\eta_2 - \nu \int_0^y \int_0^{\eta_2} u^m d\eta_1 d\eta_2 - \nu \int_0^x \int_0^{\xi_2} u^m d\xi_1 d\xi_2 \\
&+ xg_1(y) + g_2(y) + yh_1(x) + h_2(x) = \int_0^y \int_0^{\eta_2} \int_0^x \int_0^{\xi_2} f(\xi_1, \eta_1, t_m) d\xi_1 d\xi_2 d\eta_1 d\eta_2,
\end{aligned} \tag{27}$$

where $g_1(y)$, $g_2(y)$, $h_1(x)$, and $h_2(x)$ are the arbitrary functions emerged in the process of integration which can be approximated by the shifted Chebyshev polynomial interpolation. For $r \in \{1, 2\}$, define

$$h_r(x) = \sum_{i=0}^{M-1} h_r^{(i)} T_i^*(x) \text{ and } g_r(y) = \sum_{j=0}^{N-1} g_r^{(j)} T_j^*(y), \tag{28}$$

where $h_r^{(i)}$ and $g_r^{(j)}$, for $i \in \{0, 1, 2, ..., M-1\}$ and $j \in \{0, 1, 2, ..., N-1\}$, are the unknown values of these interpolated points. Next, we divide the domain Ω into a mesh with M nodes by N nodes along x-

and y-directions, respectively. We denote the nodes along the x-direction by $\mathbf{x} = \{x_1, x_2, x_3, ..., x_M\}$ and the nodes along the y-direction by $\mathbf{y} = \{y_1, y_2, y_3, ..., y_N\}$. These nodes along the x- and y-directions are the zeros of shifted Chebyshev polynomials $T_M^*(x)$ and $T_N^*(y)$, respectively. Thus, the total number of grid points in the system is $P = M \times N$, where each point is an entry in the set of Cartesian product $\mathbf{x} \times \mathbf{y}$ ordering as global type system, i.e., $(x_i, y_i) \in \mathbf{x} \times \mathbf{y}$ for $i \in \{1, 2, 3, ..., P\}$. By substituting each node in (27) and hiring $\mathbf{A}_x$ and $\mathbf{A}_y$ in Section 3.2, we can change (27) to the matrix form as

$$\sum_{j=0}^{m-1} w_j \mathbf{A}_x^2 \mathbf{A}_y^2 (\mathbf{u}^{m-j} - \mathbf{u}^{m-j-1}) + \frac{1}{2} \mathbf{A}_x \mathbf{A}_y^2 \mathrm{diag}(\mathbf{u}^{m-1}) \mathbf{u}^m + \frac{1}{2} \mathbf{A}_x^2 \mathbf{A}_y \mathrm{diag}(\mathbf{u}^{m-1}) \mathbf{u}^m$$

$$- \nu \mathbf{A}_y^2 \mathbf{u}^m - \nu \mathbf{A}_x^2 \mathbf{u}^m + \mathbf{X} \Phi_y \mathbf{g}_1 + \Phi_y \mathbf{g}_2 + \mathbf{Y} \Phi_x \mathbf{h}_1 + \Phi_y \mathbf{h}_2 = \mathbf{A}_x^2 \mathbf{A}_y^2 \mathbf{f}^m.$$

Simplifying the above equation yields

$$\mathbf{K} \mathbf{u}^m + \mathbf{X} \Phi_y \mathbf{g}_1 + \Phi_y \mathbf{g}_2 + \mathbf{Y} \Phi_x \mathbf{h}_1 + \Phi_y \mathbf{h}_2 = \mathbf{A}_x^2 \mathbf{A}_y^2 \mathbf{f}^m + w_0 \mathbf{A}_x^2 \mathbf{A}_y^2 \mathbf{u}^{m-1} - \mathbf{s}, \qquad (29)$$

where each parameter contained in (29) can be defined as follows.

$$\begin{aligned}
\mathbf{K} &= w_0 \mathbf{A}_x^2 \mathbf{A}_y^2 + \tfrac{1}{2} \mathbf{A}_x \mathbf{A}_y^2 \mathrm{diag}(\mathbf{u}^{m-1}) + \tfrac{1}{2} \mathbf{A}_x^2 \mathbf{A}_y \mathrm{diag}(\mathbf{u}^{m-1}) - \nu \mathbf{A}_y^2 - \nu \mathbf{A}_x^2, \\
\mathbf{s} &= \sum_{j=1}^{m-1} w_j \mathbf{A}_x^2 \mathbf{A}_y^2 (\mathbf{u}^{m-j} - \mathbf{u}^{m-j-1}), \\
\mathbf{X} &= \mathrm{diag}(x_1, x_2, x_3, ..., x_P), \\
\mathbf{Y} &= \mathrm{diag}(y_1, y_2, y_3, ..., y_P), \\
\mathbf{h}_r &= [h_r^{(0)}, h_r^{(1)}, h_r^{(2)}, ..., h_r^{(M-1)}]^\top \text{ for } r \in \{1, 2\}, \\
\mathbf{g}_r &= [g_r^{(0)}, g_r^{(1)}, g_r^{(2)}, ..., g_r^{(N-1)}]^\top \text{ for } r \in \{1, 2\}, \\
\mathbf{f}^m &= [f(x_1, y_1, t_m), f(x_2, y_2, t_m), f(x_3, y_3, t_m), ..., f(x_P, y_P, t_m)]^\top, \\
\mathbf{u}^m &= [u(x_1, y_1, t_m), u(x_2, y_2, t_m), u(x_3, y_3, t_m), ..., u(x_P, y_P, t_m)]^\top.
\end{aligned}$$

From (28), we obtain Φ_x and Φ_y, where

$$\Phi_x = \begin{bmatrix} T_0^*(x_1) & T_1^*(x_1) & \cdots & T_{M-1}^*(x_1) \\ T_0^*(x_2) & T_1^*(x_2) & \cdots & T_{M-1}^*(x_2) \\ \vdots & \vdots & \ddots & \vdots \\ T_0^*(x_P) & T_1^*(x_P) & \cdots & T_{M-1}^*(x_P) \end{bmatrix} \text{ and } \Phi_y = \begin{bmatrix} T_0^*(y_1) & T_1^*(y_1) & \cdots & T_{N-1}^*(y_1) \\ T_0^*(y_2) & T_1^*(y_2) & \cdots & T_{N-1}^*(y_2) \\ \vdots & \vdots & \ddots & \vdots \\ T_0^*(y_P) & T_1^*(y_P) & \cdots & T_{N-1}^*(y_P) \end{bmatrix}.$$

For the boundary conditions (24), we can transform them into the matrix form, similar the idea in [21], by employing the linear combination of the shifted Chebyshev polynomials as follows,

- Left & Right boundary conditions: For each fixed $y \in \{y_1, y_2, y_3, ..., y_N\}$, then

$$u(0, y, t_m) = \sum_{n=0}^{M-1} c_n^m T_n^*(0) := \mathbf{t}_l \mathbf{T}_M^{-1} \mathbf{u}^m(\cdot, y) = \psi_1(y, t_m) \Rightarrow (\mathbf{I}_N \otimes \mathbf{t}_l \mathbf{T}_M^{-1}) \mathbf{u}^m = \Psi_1 \qquad (30)$$

$$u(L_1, y, t_m) = \sum_{n=0}^{M-1} c_n^m T_n^*(L_1) := \mathbf{t}_r \mathbf{T}_M^{-1} \mathbf{u}^m(\cdot, y) = \psi_2(y, t_m) \Rightarrow (\mathbf{I}_N \otimes \mathbf{t}_r \mathbf{T}_M^{-1}) \mathbf{u}^m = \Psi_2 \qquad (31)$$

- Bottom & Top boundary conditions: For each fixed $x \in \{x_1, x_2, x_3, ..., x_M\}$, then

$$u(x, 0, t_m) = \sum_{n=0}^{N-1} c_n^m T_n^*(0) := \mathbf{t}_b \mathbf{T}_N^{-1} \mathbf{u}^m(x, \cdot) = \psi_3(x, t_m) \Rightarrow (\mathbf{I}_M \otimes \mathbf{t}_b \mathbf{T}_N^{-1}) \mathbf{P}^{-1} \mathbf{u}^m = \Psi_3 \qquad (32)$$

$$u(x, L_2, t_m) = \sum_{n=0}^{N-1} c_n^m T_n^*(L_2) := \mathbf{t}_t \mathbf{T}_N^{-1} \mathbf{u}^m(x, \cdot) = \psi_4(x, t_m) \Rightarrow (\mathbf{I}_M \otimes \mathbf{t}_t \mathbf{T}_N^{-1}) \mathbf{P}^{-1} \mathbf{u}^m = \Psi_4 \qquad (33)$$

where $\mathbf{I}_M$ and $\mathbf{I}_N$ are, respectively, the $M \times M$ and $N \times N$ identity matrices, $\mathbf{T}_M^{-1}$ and $\mathbf{T}_N^{-1}$ are, respectively, the $M \times M$ and $N \times N$ matrices defined in Lemma 1, $\mathbf{P}$ is defined in (6), and the other parameters are

$$\begin{aligned}
\mathbf{t}_r &= [1,1,1,...,1^{M-1}], \\
\mathbf{t}_t &= [1,1,1,...,1^{N-1}], \\
\mathbf{t}_l &= [1,-1,1,...,(-1)^{M-1}], \\
\mathbf{t}_b &= [1,-1,1,...,(-1)^{N-1}], \\
\mathbf{\Psi}_i &= [\psi_i(y_1,t_m),\psi_i(y_2,t_m),\psi_i(y_3,t_m),...,\psi_i(y_N,t_m)]^\top \text{ for } i \in \{1,2\}, \\
\mathbf{\Psi}_j &= [\psi_j(x_1,t_m),\psi_j(x_2,t_m),\psi_j(x_3,t_m),...,\psi_j(x_M,t_m)]^\top \text{ for } j \in \{3,4\}.
\end{aligned}$$

Finally, we can construct the system of iterative linear equations from Equations (29)–(33) for a total of $P + 2(M+N)$ unknowns, including $\mathbf{u}^m$, $\mathbf{g}_1$, $\mathbf{g}_2$, $\mathbf{h}_1$ and $\mathbf{h}_2$, as follows,

$$\begin{bmatrix}
\mathbf{K} & \mathbf{X}\Phi_y & \Phi_y & \mathbf{Y}\Phi_x & \Phi_x \\
\mathbf{I}_N \otimes \mathbf{t}_l\mathbf{T}_M^{-1} & 0 & 0 & \cdots & 0 \\
\mathbf{I}_N \otimes \mathbf{t}_r\mathbf{T}_M^{-1} & 0 & 0 & \cdots & 0 \\
(\mathbf{I}_M \otimes \mathbf{t}_b\mathbf{T}_N^{-1})\mathbf{P}^{-1} & \vdots & \vdots & \ddots & \vdots \\
(\mathbf{I}_M \otimes \mathbf{t}_t\mathbf{T}_N^{-1})\mathbf{P}^{-1} & 0 & 0 & \cdots & 0
\end{bmatrix}
\begin{bmatrix} \mathbf{u}^m \\ \mathbf{g}_1 \\ \mathbf{g}_2 \\ \mathbf{h}_1 \\ \mathbf{h}_2 \end{bmatrix}
=
\begin{bmatrix} \mathbf{A}_x^2\mathbf{A}_y^2(\mathbf{f}^m + w_0\mathbf{u}^{m-1}) - \mathbf{s} \\ \mathbf{\Psi}_1 \\ \mathbf{\Psi}_2 \\ \mathbf{\Psi}_3 \\ \mathbf{\Psi}_4 \end{bmatrix}. \quad (34)$$

Thus, the approximate solutions $\mathbf{u}^m$ can be reached by solving (34) in conjunction with the initial condition (23), that is, $\mathbf{u}^0 = [\phi(x_1,y_1),\phi(x_2,y_2),...,\phi(x_P,y_P)]^\top$, where for all $(x_i,y_i) \in \mathbf{x} \times \mathbf{y}$. Therefore, an arbitrary solution $u(x,y,t)$ at any fixed time t can be estimated from

$$u(x,y,t) = \mathbf{t}_y\mathbf{T}_N^{-1}(\mathbf{I}_N \otimes \mathbf{t}_x\mathbf{T}_M^{-1})\mathbf{u}^m,$$

where $\mathbf{t}_x = [T_0^*(x), T_1^*(x), T_2^*(x),..., T_{M-1}^*(x)]$ and $\mathbf{t}_y = [T_0^*(y), T_1^*(y), T_2^*(y),..., T_{N-1}^*(y)]$.

Example 3. *Consider the 2D time-fractional Burgers' Equation (22) for $(x,y) \in \Omega = (0,1) \times (0,1)$ and $t \in (0,1]$ with the forcing term*

$$f(x,y,t) = (x^2-x)(y^2-y)\left[\frac{2t^{1-\alpha}}{\Gamma(2-\alpha)} + t^2(x+y-1)(2xy-x-y)\right] - 2\nu t(x^2+y^2-x-y),$$

subject to the both homogeneous of initial and boundary conditions. The analytical solution of this problem is $u^*(x,y,t) = t(x^2-x)(y^2-y)$. For the numerical test, we pick $\nu = 100$, $\alpha = 0.5$, $\Delta t = 0.01$, and $M = N = 10$. In Table 8, the solutions approximated by our FIM-SCP Algorithm 2 are presented in the space domain Ω for various times t. We test the accuracy of our method by measuring it with the absolute error E_a. In addition, we seek the rates of convergence via ℓ^∞ norm of our Algorithm 2 with the nodal points $M = N = 10$ and different step sizes $\Delta t = 2^{-k}$ for $k \in \{4,5,6,7,8\}$, we found that these convergence rates approach to the linear convergence $O(\Delta t)$ as shown in Table 9 together with the CPU times(s). Also, the graphically numerical solutions are provided in Figure 4.

Table 8. Exact and numerical solutions of Example 3 for $\alpha = 0.5$, $M = N = 10$ and $\Delta t = 0.01$.

(x,y)	t = 0.25		t = 0.50		t = 0.75		t = 1.00	
	u(x,y,t)	E_a	u(x,y,t)	E_a	u(x,y,t)	E_a	u(x,y,t)	E_a
(0.2,0.2)	0.00641	6.73×10^{-6}	0.0128	9.52×10^{-6}	0.0192	1.17×10^{-5}	0.0256	1.35×10^{-5}
(0.4,0.4)	0.01442	1.70×10^{-5}	0.0288	2.41×10^{-5}	0.0432	2.95×10^{-5}	0.0576	3.41×10^{-5}
(0.7,0.7)	0.01104	1.25×10^{-5}	0.0221	1.77×10^{-5}	0.0331	2.16×10^{-5}	0.0441	2.50×10^{-5}
(0.9,0.9)	0.00203	1.90×10^{-6}	0.0041	2.68×10^{-6}	0.0061	3.28×10^{-6}	0.0081	3.79×10^{-6}

Algorithm 2 The numerical algorithm for solving two-dimensional time-fractional Burgers' equation

Input: $\alpha, \nu, x, y, T, M, L_1, L_2, \Delta t, \phi(x,y), \psi_1(y,t), \psi_2(y,t), \psi_3(x,t), \psi_4(x,t)$ and $f(x,y,t)$.
Output: An approximate solution $u(x,y,T)$.
1: Set $x_k = \frac{L_1}{2}\left[\cos\left(\frac{2k-1}{2M}\pi\right)+1\right]$ for $k \in \{1,2,3,...,M\}$.
2: Set $y_s = \frac{L_2}{2}\left[\cos\left(\frac{2k-1}{2N}\pi\right)+1\right]$ for $s \in \{1,2,3,...,N\}$.
3: Compute $\mathbf{X}, \mathbf{Y}, \mathbf{P}, \mathbf{t}_x, \mathbf{t}_y, \mathbf{t}_l, \mathbf{t}_r, \mathbf{t}_b, \mathbf{t}_t, \mathbf{I}_M, \mathbf{I}_N, \overline{\mathbf{T}}_M, \overline{\mathbf{T}}_N, \mathbf{T}_M^{-1}, \mathbf{T}_N^{-1}, \mathbf{A}_x, \mathbf{A}_y$ and $\mathbf{u}^0$.
4: Calculate the total number of grid points $P = M \times N$.
5: Construct x_i and y_i in the global numbering system for $i \in \{1,2,3,...,P\}$.
6: Set $t_0 = 0$ and $m = 0$.
7: **while** $t_m \leq T$ **do**
8:   Set $m = m+1$.
9:   Set $t_m = m\Delta t$.
10: Set $\mathbf{s} = \mathbf{0}$.
11: **for** $j = 1$ to $m-1$ **do**
12:    Compute $w_j = \frac{(\Delta t)^{-\alpha}}{\Gamma(2-\alpha)}[(j+1)^{1-\alpha} - j^{1-\alpha}]$.
13:    Compute $\mathbf{s} = \mathbf{s} + w_j \mathbf{A}_x^2 \mathbf{A}_y^2 (\mathbf{u}^{m-j} - \mathbf{u}^{m-j-1})$.
14: **end for**
15: Compute $\mathbf{K}, \mathbf{\Psi}_1, \mathbf{\Psi}_2, \mathbf{\Psi}_3, \mathbf{\Psi}_4$ and $\mathbf{f}^m$.
16: Find $\mathbf{u}^m$ by solving the iterative linear system (34).
17: **end while**
18: **return** $u(x,y,T) = \mathbf{t}_y \mathbf{T}_N^{-1}(\mathbf{I}_N \otimes \mathbf{t}_x \mathbf{T}_M^{-1})\mathbf{u}^m$.

Table 9. Time convergence rates and CPU time(s) for Example 3 by FIM-SCP with $M = N = 10$.

Δt	$\alpha = 0.1$			$\alpha = 0.5$			$\alpha = 0.9$		
	$\|u^* - u\|_\infty$	Rate	Time(s)	$\|u^* - u\|_\infty$	Rate	Time(s)	$\|u^* - u\|_\infty$	Rate	Time(s)
2^{-4}	3.69×10^{-3}	0.99950	0.9472	3.68×10^{-3}	0.99969	0.9694	3.68×10^{-3}	0.99994	0.9947
2^{-5}	1.83×10^{-3}	0.99950	1.3985	1.82×10^{-3}	0.99969	1.5196	1.83×10^{-3}	0.99994	1.6522
2^{-6}	8.97×10^{-4}	0.99949	3.7041	8.94×10^{-4}	0.99969	4.0292	8.96×10^{-4}	0.99994	4.4597
2^{-7}	4.32×10^{-4}	0.99947	13.718	4.29×10^{-4}	0.99968	12.710	4.32×10^{-4}	0.99994	12.606
2^{-8}	2.00×10^{-4}	0.99943	39.703	1.97×10^{-4}	0.99965	43.573	1.99×10^{-4}	0.99994	40.684

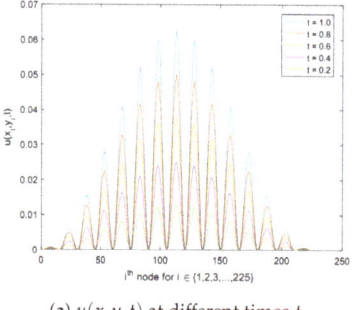

(a) $u(x,y,t)$ at different times t

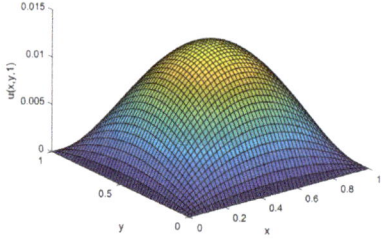

(b) Surface plot of $u(x,y,1)$

Figure 4. The graphical solutions of Example 3 for $\nu = 100$, $M = N = 15$, and $\Delta t = 0.01$.

Example 4. Consider the 2D Burgers' Equation (22) for $x \in \Omega = (0,1) \times (0,1)$ and $t \in (0,1]$ with the homogeneous initial condition and the forcing term

$$f(x,y,t) = \frac{6t^{3-\alpha}(1-x^2)^2(1-y^2)^2}{\Gamma(4-\alpha)} + 4t^6(1-x^2)^3(1-y^2)^3(x^2y + xy^2 - x - y)$$
$$- 0.4t^3\left[(y^2-1)^2(3x^2-1) + (x^2-1)^2(3y^2-1)\right],$$

subject to the boundary conditions corresponding to the analytical solution given by Cao et al. [14] is $u^*(x,y,t) = t^3(1-x^2)^2(1-y^2)^2$. By picking the parameters $\nu = 0.1$, $\alpha = 0.5$, and $M = N = 10$, the comparison of error norm L_2 between our FIM-SCP via Algorithm 2 and the discontinuous Galerkin method combined with finite different scheme (DGM-FDS) presented by Cao et al. [14] are displayed in Table 10 at time $t = 0.1$. We can see that our method gives a higher accuracy than the DGM-FDS at the same step size Δt. Next, we provide the CPU times(s) and time convergence rates based on ℓ^∞ norm of our algorithm for this problem in Table 11. Then, we see that they converge to the linear rate $O(\Delta t)$. Finally, the graphical solutions of this Example 4 are provided in Figure 5.

Table 10. Error norms L_2 between DGM-FDS and FIM-SCP of Example 4 for $M = N = 10$.

Δt	$\alpha = 0.7$		$\alpha = 0.8$		$\alpha = 0.9$	
	DGM-FDS [14]	Algorithm 2	DGM-FDS [14]	Algorithm 2	DGM-FDS [14]	Algorithm 2
0.0001	1.46×10^{-4}	3.0477×10^{-7}	1.46×10^{-4}	7.2386×10^{-7}	1.48×10^{-4}	1.6700×10^{-6}
0.00005	7.83×10^{-5}	1.2387×10^{-7}	7.76×10^{-5}	3.1525×10^{-7}	7.79×10^{-5}	7.7930×10^{-7}
0.000025	4.28×10^{-5}	5.0314×10^{-8}	4.23×10^{-5}	1.3726×10^{-7}	3.97×10^{-5}	3.6361×10^{-7}

Table 11. Time convergence rates and CPU time(s) for Example 4 by FIM-SCP with $M = N = 10$.

Δt	$\alpha = 0.1$			$\alpha = 0.5$			$\alpha = 0.9$		
	$\|u^* - u\|_\infty$	Rate	Time(s)	$\|u^* - u\|_\infty$	Rate	Time(s)	$\|u^* - u\|_\infty$	Rate	Time(s)
2^{-4}	1.76×10^{-4}	1.1426	0.9535	1.76×10^{-4}	1.1426	1.0627	1.75×10^{-4}	1.1428	1.0036
2^{-5}	9.08×10^{-5}	1.0666	2.0538	9.08×10^{-5}	1.0666	1.7050	9.06×10^{-5}	1.0667	1.6107
2^{-6}	4.61×10^{-5}	1.0323	4.3500	4.61×10^{-5}	1.0323	4.5234	4.60×10^{-5}	1.0323	3.9589
2^{-7}	2.33×10^{-5}	1.0159	12.655	2.32×10^{-5}	1.0159	12.406	2.32×10^{-5}	1.0159	11.924
2^{-8}	1.67×10^{-5}	1.0079	42.025	1.17×10^{-5}	1.0079	39.778	1.16×10^{-5}	1.0079	41.899

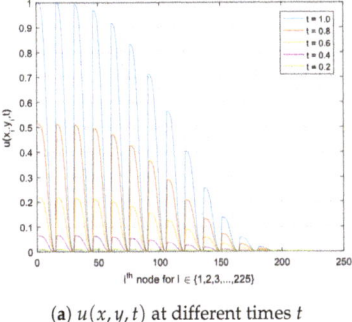
(a) $u(x,y,t)$ at different times t

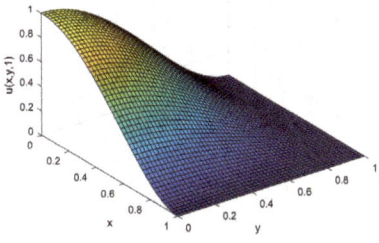
(b) Surface plot of $u(x,y,1)$

Figure 5. The graphical solutions of Example 4 for $\nu = 0.1$, $M = N = 15$, and $\Delta t = 0.01$.

4.3. Algorithm for Time-Fractional Coupled Burgers' Equation

Consider the following coupled Burgers' equation with fractional time derivative for $\alpha \in (0,1]$

$$\frac{\partial^\alpha u}{\partial t^\alpha} = \frac{\partial^2 u}{\partial x^2} + 2u\frac{\partial u}{\partial x} - \frac{\partial(uv)}{\partial x} + f(x,t), \ x \in (0,L), \ t \in (0,T]$$
$$\frac{\partial^\beta v}{\partial t^\beta} = \frac{\partial^2 v}{\partial x^2} + 2v\frac{\partial v}{\partial x} - \frac{\partial(uv)}{\partial x} + g(x,t), \ x \in (0,L), \ t \in (0,T]$$
(35)

subject to the initial conditions
$$u(x,0) = \phi_1(x), \ x \in [0,L],$$
$$v(x,0) = \phi_2(x), \ x \in [0,L],$$
(36)

and the boundary conditions
$$u(0,t) = \psi_1(t), \ u(L,t) = \psi_2(t), \ t \in (0,T],$$
$$v(0,t) = \psi_3(t), \ v(L,t) = \psi_4(t), \ t \in (0,T],$$
(37)

where $f(x,t)$, $g(x,t)$, $\phi_1(x)$, $\phi_2(x)$, $\varphi_1(t)$, $\varphi_2(t)$, $\varphi_3(t)$, and $\varphi_4(t)$ are the given functions. The procedure of using our FIM for solving u and v are similar, we only discuss here the details in finding the approximate solution u.

We begin with linearizing the system (35) by taking the an iteration of time $t_m = m(\Delta t)$ for $m \in \mathbb{N}$, where Δt is a time step. We obtain

$$\left.\frac{\partial^\alpha u}{\partial t^\alpha}\right|_{t=t_m} = \frac{\partial^2 u^m}{\partial x^2} + 2u^{m-1}\frac{\partial u^m}{\partial x} - \frac{\partial(v^{m-1}u^m)}{\partial x} + f(x,t_m),$$
$$\left.\frac{\partial^\beta v}{\partial t^\beta}\right|_{t=t_m} = \frac{\partial^2 v^m}{\partial x^2} + 2v^{m-1}\frac{\partial v^m}{\partial x} - \frac{\partial(u^{m-1}v^m)}{\partial x} + g(x,t_m),$$

where $u^m = u(x,t_m)$ and $v^m = v(x,t_m)$ are numerical solutions of u and v in the mth iteration, respectively. Next, let us consider the fractional time derivative for $\alpha \in (0,1]$ in the Caputo sense by using the same procedure as in (12), by taking the double layer integration on both sides, we obtain

$$\sum_{j=0}^{m-1} w_j^\alpha \int_0^{x_k}\int_0^\eta (u^{m-j} - u^{m-j-1})d\xi d\eta = u^m(x_k) + 2\int_0^{x_k}\int_0^\eta \left(u^{m-1}\frac{\partial u^m}{\partial \xi}\right)d\xi d\eta$$
$$- \int_0^{x_k}(v^{m-1}u^m)d\eta + \int_0^{x_k}\int_0^\eta f(\xi,t_m)d\xi d\eta + d_1 x_k + d_2,$$
(38)

$$\sum_{j=0}^{m-1} w_j^\beta \int_0^{x_k}\int_0^\eta (v^{m-j} - v^{m-j-1})d\xi d\eta = v^m(x_k) + 2\int_0^{x_k}\int_0^\eta \left(v^{m-1}\frac{\partial v^m}{\partial \xi}\right)d\xi d\eta$$
$$- \int_0^{x_k}(u^{m-1}v^m)d\eta + \int_0^{x_k}\int_0^\eta g(\xi,t_m)d\xi d\eta + d_3 x_k + d_4,$$
(39)

where $w_j^\gamma = \frac{(\Delta t)^{-\gamma}}{\Gamma(2-\gamma)}\left[(j+1)^{1-\gamma} - j^{1-\gamma}\right]$ for $\gamma \in \{\alpha, \beta\}$, and $d_1, d_2, d_3,$ and d_4 are arbitrary constants of integration. For the nonlinear terms in (38) and (39), by using the same process as in (15), we let

$$q_1(x_k) := \int_0^{x_k}\int_0^\eta \left(u^{m-1}\frac{\partial u^m}{\partial \xi}\right)d\xi d\eta = \int_0^{x_k} u^{m-1}u^m d\eta - \int_0^{x_k}\int_0^\eta \mathbf{T}'(\xi)\mathbf{T}^{-1}\mathbf{u}^{m-1}u^m d\xi d\eta,$$

$$q_2(x_k) := \int_0^{x_k}\int_0^\eta \left(v^{m-1}\frac{\partial v^m}{\partial \xi}\right)d\xi d\eta = \int_0^{x_k} v^{m-1}v^m d\eta - \int_0^{x_k}\int_0^\eta \mathbf{T}'(\xi)\mathbf{T}^{-1}\mathbf{v}^{m-1}v^m d\xi d\eta.$$

For computational convenience, we express $q_1(x_k)$ and $q_2(x_k)$ into matrix forms as

$$\mathbf{q}_1 = \mathbf{A}\operatorname{diag}(\mathbf{u}^{m-1})\mathbf{u}^m - \mathbf{A}^2\operatorname{diag}(\mathbf{T}'\mathbf{T}^{-1}\mathbf{u}^{m-1})\mathbf{u}^m := \mathbf{Q}_1 \mathbf{u}^m, \tag{40}$$

$$\mathbf{q}_2 = \mathbf{A}\operatorname{diag}(\mathbf{v}^{m-1})\mathbf{v}^m - \mathbf{A}^2\operatorname{diag}(\mathbf{T}'\mathbf{T}^{-1}\mathbf{v}^{m-1})\mathbf{v}^m := \mathbf{Q}_2 \mathbf{v}^m, \tag{41}$$

where $\mathbf{T}'$ is defined in (17) and other parameters obtained on (40) and (41) are

$$
\begin{aligned}
\mathbf{Q}_1 &= \mathbf{A}\operatorname{diag}(\mathbf{u}^{m-1}) - \mathbf{A}^2\operatorname{diag}(\mathbf{T}'\mathbf{T}^{-1}\mathbf{u}^{m-1}), \\
\mathbf{Q}_2 &= \mathbf{A}\operatorname{diag}(\mathbf{v}^{m-1}) - \mathbf{A}^2\operatorname{diag}(\mathbf{T}'\mathbf{T}^{-1}\mathbf{v}^{m-1}), \\
\mathbf{u}^m &= [u(x_1,t_m), u(x_2,t_m), u(x_3,t_m), ..., u(x_M,t_m)]^\top, \\
\mathbf{v}^m &= [v(x_1,t_m), v(x_2,t_m), v(x_3,t_m), ..., v(x_M,t_m)]^\top, \\
\mathbf{q}_i &= [q_i(x_1), q_i(x_2), q_i(x_3), ..., q_i(x_M)]^\top \text{ for } i \in \{1,2\}.
\end{aligned}
$$

Consequently, using (40), (41), and the procedure in Section 3.1, we can convert both (38) and (39) into the matrix forms as

$$\sum_{j=0}^{m-1} w_j^\alpha \mathbf{A}^2 (\mathbf{u}^{m-j} - \mathbf{u}^{m-j-1}) = \mathbf{u}^m + 2\mathbf{Q}_1 \mathbf{u}^m - \mathbf{A}\operatorname{diag}(\mathbf{v}^{m-1})\mathbf{u}^m + \mathbf{A}^2 \mathbf{f}^m + d_1 \mathbf{x} + d_2 \mathbf{i},$$

$$\sum_{j=0}^{m-1} w_j^\beta \mathbf{A}^2 (\mathbf{v}^{m-j} - \mathbf{v}^{m-j-1}) = \mathbf{v}^m + 2\mathbf{Q}_2 \mathbf{v}^m - \mathbf{A}\operatorname{diag}(\mathbf{u}^{m-1})\mathbf{v}^m + \mathbf{A}^2 \mathbf{g}^m + d_3 \mathbf{x} + d_4 \mathbf{i}.$$

Rearranging the above system yields

$$\left[\mathbf{I} + 2\mathbf{Q}_1 - \mathbf{A}\operatorname{diag}(\mathbf{v}^{m-1}) - w_0^\alpha \mathbf{A}^2\right] \mathbf{u}^m + d_1 \mathbf{x} + d_2 \mathbf{i} = \mathbf{s}_1 - w_0^\alpha \mathbf{A}^2 \mathbf{u}^{m-1} - \mathbf{A}^2 \mathbf{f}^m, \tag{42}$$

$$\left[\mathbf{I} + 2\mathbf{Q}_2 - \mathbf{A}\operatorname{diag}(\mathbf{u}^{m-1}) - w_0^\beta \mathbf{A}^2\right] \mathbf{v}^m + d_3 \mathbf{x} + d_4 \mathbf{i} = \mathbf{s}_2 - w_0^\beta \mathbf{A}^2 \mathbf{v}^{m-1} - \mathbf{A}^2 \mathbf{g}^m, \tag{43}$$

where $\mathbf{I}$ is the $M \times M$ identity matrix and other parameters are defined by

$$
\begin{aligned}
\mathbf{s}_1 &= \sum_{j=1}^{m-1} w_j^\alpha \mathbf{A}^2 (\mathbf{u}^{m-j} - \mathbf{u}^{m-j-1}), \\
\mathbf{s}_2 &= \sum_{j=1}^{m-1} w_j^\beta \mathbf{A}^2 (\mathbf{v}^{m-j} - \mathbf{v}^{m-j-1}), \\
\mathbf{f}^m &= [f(x_1,t_m), f(x_2,t_m), f(x_3,t_m), ..., f(x_M,t_m)]^\top, \\
\mathbf{g}^m &= [g(x_1,t_m), g(x_2,t_m), g(x_3,t_m), ..., g(x_M,t_m)]^\top.
\end{aligned}
$$

The boundary conditions (37) are transformed into the vector forms by using the same process as in (19) and (20), that is,

$$\mathbf{t}_l \mathbf{T}^{-1} \mathbf{u}^m = \psi_1(t_m) \text{ and } \mathbf{t}_r \mathbf{T}^{-1} \mathbf{u}^m = \psi_2(t_m), \tag{44}$$

$$\mathbf{t}_l \mathbf{T}^{-1} \mathbf{v}^m = \psi_3(t_m) \text{ and } \mathbf{t}_r \mathbf{T}^{-1} \mathbf{v}^m = \psi_4(t_m), \tag{45}$$

where $\mathbf{t}_l = [1, -1, 1, ..., (-1)^{M-1}]$ and $\mathbf{t}_r = [1, 1, 1, ..., 1]$. Finally, starting from the initial guesses

$$\mathbf{u}^0 = [\phi_1(x_1), \phi_1(x_2), \phi_1(x_3), ..., \phi_1(x_M)]^\top \text{ and } \mathbf{v}^0 = [\phi_2(x_1), \phi_2(x_2), \phi_2(x_3), ..., \phi_2(x_M)]^\top,$$

we can construct the system of the mth iterative linear equations for finding numerical solutions. The approximate solutions of u can be obtained from (42) and (44) while the approximate solutions of v can be reached by using (43) and (45):

$$\begin{bmatrix} \mathbf{I} + 2\mathbf{Q}_1 - \mathbf{A}\operatorname{diag}(\mathbf{v}^{m-1}) - w_0^\alpha \mathbf{A}^2 & \mathbf{x} & \mathbf{i} \\ \mathbf{t}_l \mathbf{T}^{-1} & 0 & 0 \\ \mathbf{t}_r \mathbf{T}^{-1} & 0 & 0 \end{bmatrix} \begin{bmatrix} \mathbf{u}^m \\ d_1 \\ d_2 \end{bmatrix} = \begin{bmatrix} \mathbf{s}_1 - w_0^\alpha \mathbf{A}^2 \mathbf{u}^{m-1} - \mathbf{A}^2 \mathbf{f}^m \\ \psi_1(t_m) \\ \psi_2(t_m) \end{bmatrix}, \tag{46}$$

and

$$\begin{bmatrix} \mathbf{I} + 2\mathbf{Q}_2 - \mathbf{A}\mathrm{diag}(\mathbf{u}^{m-1}) - w_0^\beta \mathbf{A}^2 & \mathbf{x} & \mathbf{i} \\ \mathbf{t}_l \mathbf{T}^{-1} & 0 & 0 \\ \mathbf{t}_r \mathbf{T}^{-1} & 0 & 0 \end{bmatrix} \begin{bmatrix} \mathbf{v}^m \\ d_3 \\ d_4 \end{bmatrix} = \begin{bmatrix} \mathbf{s}_2 - w_0^\beta \mathbf{A}^2 \mathbf{v}^{m-1} - \mathbf{A}^2 \mathbf{g}^m \\ \psi_3(t_m) \\ \psi_4(t_m) \end{bmatrix}. \quad (47)$$

For any fixed t, the approximate solutions of $u(x,t)$ and $v(x,t)$ on the space domain can be obtained by computing $u(x,t) = \mathbf{t}_x \mathbf{T}^{-1} \mathbf{u}^m$ and $v(x,t) = \mathbf{t}_x \mathbf{T}^{-1} \mathbf{v}^m$, where $\mathbf{t}_x = [T_0^*(x), T_1^*(x), T_2^*(x), ..., T_{M-1}^*(x)]$.

Example 5. *Consider the time-fractional coupled Burgers' Equation (35) for $x \in (0,1)$ and $t \in (0,1]$ with the forcing terms*

$$f(x,t) = \frac{6xt^{3-\alpha}}{\Gamma(4-\alpha)} \text{ and } g(x,t) = \frac{6xt^{3-\beta}}{\Gamma(4-\beta)}$$

subject to the homogeneous initial conditions and the boundary conditions corresponding to the analytical solution given by Albuohimad and Adibi [25] is $u^*(x,t) = v^*(x,t) = xt^3$. For the numerical test, we choose the kinematic viscosity $\nu = 1$, $\alpha = \beta = 0.5$ and $M = 40$. Table 12 presents the exact solution $u^*(x,1)$ and the numerical solutions $u(x,1)$ together with $v(x,1)$ by using our FIM-SCP through Algorithm 3. The accuracy is measured by the absolute error E_a. Table 13 displays the comparison of the error norms L_∞ of our approximate solutions and the approximate solutions obtained by using the collocation method with FDM (CM-FDM) introduced by Albuohimad and Adibi in [25]. As can be seen from Table 13, our FIM-SCP Algorithm 3 is more accurate. Next, the time convergence rates based on ℓ^∞ and CPU times(s) of this problem that solved by Algorithm 3 are demonstrated in Table 14. Since the approximate solutions u and v are the same, we only present the graphical solution of u in Figure 6.

Algorithm 3 The numerical algorithm for solving 1D time-fractional coupled Burgers' equation

---

Input: $\alpha, \beta, x, L, T, M, \Delta t, \phi_1(x), \phi_2(x), \psi_1(t), \psi_2(t), \psi_3(t), \psi_4(t), f(x,t)$ and $g(x,t)$.
Output: The approximate solutions $u(x,T)$ and $v(x,T)$.
1: Set $x_k = \frac{L}{2}\left[\cos\left(\frac{2k-1}{2M}\pi\right) + 1\right]$ for $k \in \{1,2,3,...,M\}$.
2: Compute $\mathbf{x}, \mathbf{i}, \mathbf{t}_l, \mathbf{t}_r, \mathbf{t}_x, \mathbf{A}, \mathbf{I}, \mathbf{T}, \mathbf{T}', \overline{\mathbf{T}}, \mathbf{T}^{-1}, \mathbf{u}^0$ and $\mathbf{v}^0$.
3: Set $t_0 = 0$ and $m = 0$.
4: **while** $t_m \leq T$ **do**
5:     Set $m = m + 1$.
6:     Set $t_m = m\Delta t$.
7:     Set $\mathbf{s}_1 = \mathbf{0}$ and $\mathbf{s}_2 = \mathbf{0}$.
8:     **for** $j = 1$ to $m-1$ **do**
9:         Compute $w_j^\alpha = \frac{(\Delta t)^{-\alpha}}{\Gamma(2-\alpha)}[(j+1)^{1-\alpha} - j^{1-\alpha}]$.
10:        Compute $w_j^\beta = \frac{(\Delta t)^{-\beta}}{\Gamma(2-\beta)}[(j+1)^{1-\beta} - j^{1-\beta}]$.
11:        Compute $\mathbf{s}_1 = \mathbf{s}_1 + w_j^\alpha \mathbf{A}^2(\mathbf{u}^{m-j} - \mathbf{u}^{m-j-1})$.
12:        Compute $\mathbf{s}_2 = \mathbf{s}_2 + w_j^\beta \mathbf{A}^2(\mathbf{v}^{m-j} - \mathbf{v}^{m-j-1})$.
13:     **end for**
14:     Calculate $\mathbf{Q}_1, \mathbf{Q}_2, \mathbf{f}^m$ and $\mathbf{g}^m$.
15:     Find $\mathbf{u}^m$ by solving the iterative linear system (46).
16:     Find $\mathbf{v}^m$ by solving the iterative linear system (47).
17: **end while**
18: **return** $u(x,T) = \mathbf{t}_x(\mathbf{T}^*)^{-1}\mathbf{u}^m$ and $v(x,T) = \mathbf{t}_x(\mathbf{T}^*)^{-1}\mathbf{v}^m$.

Table 12. Comparison of exact and numerical solutions of Example 5 for $\alpha = \beta = 0.5$, $M = 40$.

Δt	x	$u^*(x,1)$	$u(x,1)$	$E_a(u)$	$v(x,1)$	$E_a(v)$
0.005	0.2	0.2	0.200014	1.3637×10^{-5}	0.200014	1.3637×10^{-5}
	0.4	0.4	0.400024	2.4030×10^{-5}	0.400024	2.4030×10^{-5}
	0.6	0.6	0.600028	2.7782×10^{-5}	0.600028	2.7782×10^{-5}
0.001	0.2	0.2	0.200001	1.2398×10^{-6}	0.200001	1.2398×10^{-6}
	0.4	0.4	0.400002	2.1845×10^{-6}	0.400002	2.1845×10^{-6}
	0.6	0.6	0.600003	2.5250×10^{-6}	0.600003	2.5250×10^{-6}
0.0005	0.2	0.2	0.200000	4.4002×10^{-7}	0.200000	4.4002×10^{-7}
	0.4	0.4	0.400001	7.7529×10^{-7}	0.400001	7.7529×10^{-7}
	0.6	0.6	0.600001	8.9611×10^{-7}	0.600001	8.9611×10^{-7}

Table 13. Comparison of error norms L_∞ of Example 5 for $\alpha = \beta = 0.5$, $M = 5$ and $t = 1$.

Δt	CM-FDM [25]		FIM-SCP Algorithm 3	
	$L_\infty(u)$	$L_\infty(v)$	$L_\infty(u)$	$L_\infty(v)$
0.03125	$3.96243489 \times 10^{-4}$	$3.96243489 \times 10^{-4}$	2.0275×10^{-4}	2.0275×10^{-4}
0.015625	$1.46199451 \times 10^{-4}$	$1.46199451 \times 10^{-4}$	7.3260×10^{-5}	7.3260×10^{-5}
0.0078125	$5.30198057 \times 10^{-5}$	$5.30198057 \times 10^{-5}$	2.6297×10^{-5}	2.6297×10^{-5}
0.00390625	$1.90424033 \times 10^{-5}$	$1.90424033 \times 10^{-5}$	9.3967×10^{-6}	9.3967×10^{-6}
0.001953125	$6.80038150 \times 10^{-6}$	$6.80038150 \times 10^{-6}$	3.3472×10^{-6}	3.3472×10^{-6}

Table 14. Time convergence rates and CPU time(s) for Example 5 by FIM-SCP with $M = 20$.

Δt	$\alpha = \beta = 0.1$			$\alpha = \beta = 0.5$			$\alpha = \beta = 0.9$		
	$\|u^* - u\|_\infty$	Rate	Time(s)	$\|u^* - u\|_\infty$	Rate	Time(s)	$\|u^* - u\|_\infty$	Rate	Time(s)
2^{-4}	1.41×10^{-3}	1.1426	0.3901	1.41×10^{-3}	1.1426	0.4008	1.40×10^{-3}	1.1427	0.4801
2^{-5}	7.26×10^{-4}	1.0666	0.4064	7.26×10^{-4}	1.0666	0.4292	7.25×10^{-4}	1.0667	0.4895
2^{-6}	3.69×10^{-4}	1.0323	0.8505	3.69×10^{-4}	1.0323	0.9028	3.68×10^{-4}	1.0323	0.8715
2^{-7}	1.86×10^{-4}	1.0159	2.5623	1.86×10^{-4}	1.0159	2.4748	1.86×10^{-4}	1.0159	2.7062
2^{-8}	9.32×10^{-5}	1.0079	8.8157	9.32×10^{-5}	1.0079	8.2575	9.32×10^{-5}	1.0079	8.4962

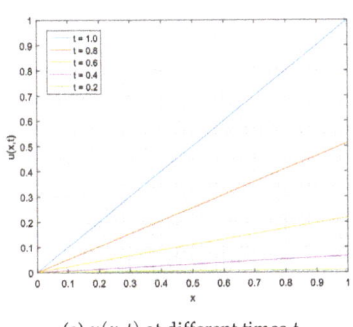

(a) $u(x,t)$ at different times t

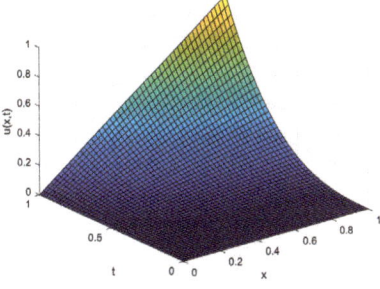

(b) Surface plot of $u(x,t)$

Figure 6. The graphical solutions of Example 5 for $\alpha = \beta = 0.5$, $M = 40$, and $\Delta t = 0.001$.

Example 6. *Consider the time-fractional coupled Burgers' Equation (35) for $x \in (0,1)$ and $t \in (0,1]$ with the forcing terms*

$$f(x,t) = \left[\frac{\Gamma(4)t^{-\alpha}}{\Gamma(4-\alpha)} + 1\right] t^3 \sin(x) \quad \text{and} \quad g(x,t) = \left[\frac{\Gamma(4)t^{-\beta}}{\Gamma(4-\beta)} + 1\right] t^3 \sin(x)$$

subject to the homogeneous initial conditions and the boundary conditions corresponding to the analytical solution given by Albuohimad and Adibi [25] is $u^*(x,t) = v^*(x,t) = t^3 \sin(x)$. For the numerical test, we choose the viscosity $\nu = 1$, $\alpha = \beta = 0.5$ and $M = 5$. Table 15 provides the comparison of error norms L_∞ between our FIM-SCP and the CM-FDM in [25] for various values of Δt and M, it show that our method is more accurate. Moreover, Table 16 illustrates the rates of convergence and CPU times(s) for $M = 20$. Figure 7a,b show the numerical solutions $u(x,t)$ at different times t and the surface plot of $u(x,t)$, respectively. Note that we only show the graphical solution of $u(x,t)$ since the approximate solutions $u(x,t)$ and $v(x,t)$ are the same.

Table 15. Comparison of error norms L_∞ between CM-FDM and FIM-SCP for Example 6.

M	Δt	CM-FDM [25]		FIM-SCP Algorithm 3	
		$L_\infty(u)$	$L_\infty(v)$	$L_\infty(u)$	$L_\infty(v)$
5	1/4	$2.38860019 \times 10^{-3}$	$2.38860019 \times 10^{-3}$	1.3600×10^{-3}	1.3600×10^{-3}
5	1/16	$3.68124891 \times 10^{-4}$	$3.68124891 \times 10^{-4}$	1.5995×10^{-4}	1.5995×10^{-4}
5	1/32	$1.33717524 \times 10^{-4}$	$1.33717524 \times 10^{-4}$	5.3813×10^{-5}	5.3813×10^{-5}
3	1/128	$2.16075055 \times 10^{-3}$	$2.16075055 \times 10^{-3}$	2.7726×10^{-3}	2.7726×10^{-3}
4	1/128	$1.41457658 \times 10^{-4}$	$1.41457658 \times 10^{-4}$	1.6397×10^{-4}	1.6397×10^{-4}
5	1/128	$4.69272546 \times 10^{-5}$	$4.69272546 \times 10^{-5}$	1.7565×10^{-5}	1.7565×10^{-5}

Table 16. Time convergence rates and CPU time(s) for Example 6 by FIM-SCP with $M = 20$.

Δt	$\alpha = \beta = 0.1$			$\alpha = \beta = 0.5$			$\alpha = \beta = 0.9$		
	$\|u^* - u\|_\infty$	Rate	Time(s)	$\|u^* - u\|_\infty$	Rate	Time(s)	$\|u^* - u\|_\infty$	Rate	Time(s)
2^{-4}	1.18×10^{-3}	1.1427	0.4041	1.18×10^{-3}	1.1427	0.3873	1.18×10^{-3}	1.1427	0.3982
2^{-5}	6.11×10^{-4}	1.0667	0.4902	6.11×10^{-4}	1.0667	0.4468	6.11×10^{-4}	1.0667	0.4245
2^{-6}	3.11×10^{-4}	1.0323	0.8941	3.11×10^{-4}	1.0323	0.8829	3.11×10^{-4}	1.0323	0.8873
2^{-7}	1.57×10^{-4}	1.0159	2.5981	1.57×10^{-4}	1.0159	2.6828	1.57×10^{-4}	1.0159	2.4627
2^{-8}	7.91×10^{-5}	1.0079	7.9922	7.91×10^{-5}	1.0079	8.3994	7.90×10^{-5}	1.0079	8.2681

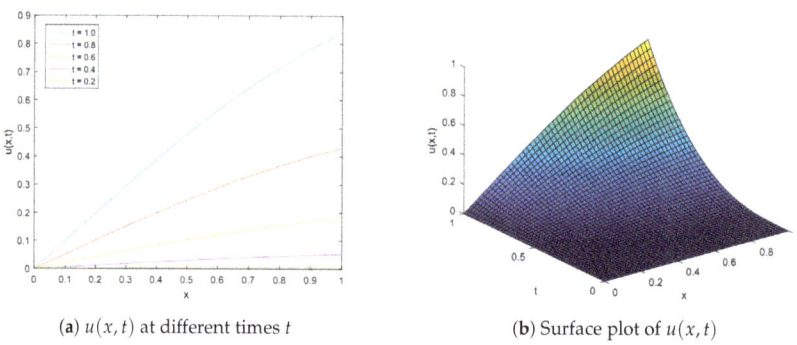

(a) $u(x,t)$ at different times t (b) Surface plot of $u(x,t)$

Figure 7. The graphical results of Example 6 for $\alpha = \beta = 0.5$, $M = 40$ and $\Delta t = 0.001$.

5. Conclusions and Discussion

In this paper, we applied our improved FIM-SCP to develop the decent and accurate numerical algorithms for finding the approximate solutions of time-fractional Burgers' equations both in one- and two-dimensional spatial domains and time-fractional coupled Burgers' equations. Their fractional-order derivatives with respect to time were described in the Caputo sense and estimated by forward difference quotient. According to Example 1, even though, we obtain similar accuracy, however, it can be seen that our method does not require the solution to be separable among the spatial and temporal variables. For Example 2, the results confirm that even with nonlinear FDEs, the FIM-SCP provides better accuracy than FDM. For two dimensions, Example 4 shows that even with the small kinematic viscosity ν, our method can deal with a shock wave solution, which is not globally continuously differentiable as that of the classical Burgers' equation under the same effect of small kinematic viscosity ν. We can also see from Examples 5 and 6 that our proposed method can be extended to solve the time-fractional Burgers' equation and we expect that it will also credibly work with other system of time-fractional nonlinear equation. We notice that our method provides better accuracy even when we use a small number of nodal points. Evidently, when we decrease the time step, it furnishes more accurate results. Also, we illustrated the time convergence rate of our method based on ℓ^∞ norm, we observe that it approaches to the linear convergence $O(\Delta t)$. Finally, we show the computational cost in terms of CPU time(s) for each example. An interesting direction for our future work is to extend our technique to solve space-fractional Burgers' equations and other nonlinear FDEs.

Author Contributions: Conceptualization, A.D., R.B., and T.T.; methodology, R.B.; software, A.D.; validation, A.D., R.B. and T.T.; formal analysis, R.B.; investigation, A.D.; writing—original draft preparation, A.D.; writing—review and editing, R.B. and T.T.; visualization, A.D.; supervision, R.B. and T.T.; project administration, R.B.; funding acquisition, A.D.

Funding: This research was funded by The 100th Anniversary Chulalongkorn University Fund for Doctoral Scholarship.

Conflicts of Interest: The authors declare no conflicts of interest.

Abbreviations

The following abbreviations are used in this manuscript.

CM-FDM	collocation method with finite difference method
DGM-FDS	discontinuous Galerkin method with finite different scheme
EPM-CHT	expansion method with Cole–Hopf transformation
FDE	fractional differential equation
FDM	finite difference method
FIM	finite integration method
FIM-SCP	finite integration method with shifted Chebyshev polynomial
PDE	partial differential equation
QBS-FEM	quadratic B-spline finite element Galerkin method

References

1. Metzler, R.; Klafter, J. The random walk's guide to anomalous diffusion: A fractional dynamics approach. *Phys. Rep.* **2000**, *339*, 1–77. [CrossRef]
2. Kumar, D.; Singh, J.; Baleanu, D. A fractional model of convective radial fins with temperature-dependent thermal conductivity. *Rom. Rep. Phys.* **2017**, *69*, 1–13.
3. Agila, A.; Baleanu, D.; Eid, R.; Irfanoglu, B. Applications of the extended fractional Euler-Lagrange equations model to freely oscillating dynamical system. *Rom. J. Phys.* **2016**, *61*, 350–359.
4. Yang, X.J.; Gao, F.; Srivastava, H.M. New rheological models within local fractional derivative. *Rom. Rep. Phys.* **2017**, *69*, 1–12.
5. Laskin, N. Principles of fractional quantum mechanics. In *Fractional Dynamics*; World Scientific: Singapore, 2011; pp. 393–427.
6. Burgers, J.M. A mathematical model illustrating the theory of turbulence. *Adv. Appl. Mech.* **1948**, *1*, 171–199.
7. Esen, A.; Yagmurlu, N.M.; Tasbozan, O. Approximate analytical solution to time-fractional damped Burgers and Cahn-Allen equations. *Appl. Math. Inf. Sci.* **2013**, *7*, 1951–1956. [CrossRef]

8. Su, N.; Watt, J.; Vincent, K.W.; Close, M.E.; Mao, R. Analysis of turbulent flow patterns of soil water under field conditions using Burgers' equation and porous suction-cup samplers. *Aust. J. Soil Res.* **2004**, *42*, 9–16. [CrossRef]
9. Zhang, Y.; Baleanu, D.; Yang, X.J. New solutions of the transport equations in porous media within local fractional derivative. *Proc. Rom. Acad. Ser. A Math. Phys. Tech. Sci. Inf. Sci.* **2016**, *17*, 230–236.
10. Wang, Q. Numerical solutions for fractional KdV-Burgers equation by Adomian decomposition method. *Appl. Math. Comput.* **2006**, *182*, 1048–1055.
11. Guesmia, A.; Daili, N. Numerical approximation of fractional Burgers equation. *Commun. Math. Appl.* **2010**, *1*, 77–90.
12. Esen, A.; Tasbozan, O. Numerical solution of time fractional Burgers equation. *Acta Univ. Sapientiae Math.* **2015**, *7*, 167–185. [CrossRef]
13. Yokus, A.; Kaya, D. Numerical and exact solutions for time fractional Burgers' equation. *J. Nonlinear Sci. Appl.* **2017**, *10*, 3419–3428. [CrossRef]
14. Cao, W.; Xu, Q.; Zheng, Z. Solution of two-dimensional time-fractional Burgers equation with high and low Reynolds numbers. *Adv. Differ. Equ.* **2017**, *338*, 1–14. [CrossRef]
15. Wen, P.; Hon, Y.; Li, M.; Korakianitis, T. Finite integration method for partial differential equations. *Appl. Math. Model.* **2013**, *37*, 10092–10106. [CrossRef]
16. Li, M.; Chen, C.; Hon, Y.; Wen, P. Finite integration method for solving multi-dimensional partial differential equations. *Appl. Math. Model.* **2015**, *39*, 4979–4994. [CrossRef]
17. Li, M.; Tian, Z.; Hon, Y.; Chen, C.; Wen, P. Improved Finite integration method for partial differential equations. *Eng. Anal. Bound. Elem.* **2016**, *64*, 230–236. [CrossRef]
18. Li, Y.; Hon, Y.L. Finite integration method with radial basis function for solving stiff problems. *Eng. Anal. Bound. Elem.* **2017**, *82*, 32–42.
19. Li, Y.; Li, M.; Hon, Y.C. Improved finite integration method for multi-dimensional nonlinear Burgers' equation with shock wave. *Neural Parallel Sci. Comput.* **2015**, *23*, 63–86.
20. Yun, D.L.; Wen, Z.L.; Hon, Y.L. Adaptive least squares finite integration method for higher-dimensional singular perturbation problems with multiple boundary layers. *Appl. Math. Comput.* **2015**, *271*, 232–250. [CrossRef]
21. Boonklurb, R.; Duangpan, A.; Treeyaprasert, T. Modified finite integration method using Chebyshev polynomial for solving linear differential equations. *J. Numer. Ind. Appl. Math.* **2018**, *12*, 1–19.
22. Rivlin, T.J. *Chebyshev Polynomials, From Approximation Theory to Algebra and Number Theory*, 2nd ed.; John Wiley & Sons: New York, NY, USA, 1990.
23. Podlubny, I. *Fractional Differential Equations*; Academic Press: San Diego, CA, USA, 1999.
24. Zhang, H.; Ding, F. On the Kronecker products and their applications. *J. Appl. Math.* **2013**, *2013*, 296185. [CrossRef]
25. Albuohimad, A.; Adibi, H. The Chebyshev collocation solution of the time fractional coupled Burgers' equation. *J. Math. Comput. Sci.* **2017**, *17*, 179–193. [CrossRef]

© 2019 by the authors. Licensee MDPI, Basel, Switzerland. This article is an open access article distributed under the terms and conditions of the Creative Commons Attribution (CC BY) license (http://creativecommons.org/licenses/by/4.0/).

Article

Orhonormal Wavelet Bases on The 3D Ball Via Volume Preserving Map from The Regular Octahedron

Adrian Holhoş and Daniela Roşca *

Department of Mathematics, Technical University of Cluj-Napoca, str. Memorandumului 28, RO-400114 Cluj-Napoca, Romania; Adrian.Holhos@math.utcluj.ro
* Correspondence: Daniela.Rosca@math.utcluj.ro

Received: 13 May 2020; Accepted: 12 June 2020; Published: 17 June 2020

Abstract: We construct a new volume preserving map from the unit ball $\mathbb{B}^3$ to the regular 3D octahedron, both centered at the origin, and its inverse. This map will help us to construct refinable grids of the 3D ball, consisting in diameter bounded cells having the same volume. On this 3D uniform grid, we construct a multiresolution analysis and orthonormal wavelet bases of $L^2(\mathbb{B}^3)$, consisting in piecewise constant functions with small local support.

Keywords: wavelets on 3D ball; uniform 3D grid; volume preserving map

1. Introduction

Spherical 3D signals occur in a wide range of fields, including computer graphics, and medical imaging (e.g., 3D reconstruction of medical images [1]), crystallography (texture analysis of crystals) [2,3] and geoscience [4–6]. Therefore, we need suitable efficient techniques for manipulating such signals, and one of the most efficient technique consists in using wavelets on the 3D ball (see e.g., [4–10] and the references therein). In this paper we propose to construct an orthonormal basis of wavelets with small support, defined on the 3D ball $\mathbb{B}^3$, starting from a multiresolution analysis. Our wavelets will be piecewise constant functions on the cells of a uniform and refinable grid of $\mathbb{B}^3$. By a refinable (or hierarchical) grid we mean that the cells can be divided successively into a given number of smaller cells of the same volume. By a uniform grid we mean that all the cells at a certain level of subdivision have the same volume. These two very important properties of our grid derive from the fact that it is constructed by mapping a uniform and refinable grid of the 3D regular octahedron, using a volume preserving map onto $\mathbb{B}^3$. Compared to the wavelets on the 3D ball constructed in [8,10], with localized support, our wavelets have local support, and this is very important when dealing with data consisting in big jumps on small portions, as shown in [11]. Another construction of piecewise constant wavelets on the 3D ball was realized in [7], starting from a similar construction on the 2D sphere. The author assumes that his wavelets are the first Haar wavelets on the 3D ball which are orthogonal and symmetric, even though we do not see any symmetry, neither in the cells, nor in the decomposition matrix. Moreover, his 8×8 decomposition matrices change in each step of the refinement, the entries depending on the volumes of the cells, which are, in our opinion, difficult to evaluate and for this reason they are not calculated explicitly in [7]. Another advantage of our construction is that our cells are diameter bounded, unlike the cells in [7] containing the origin, which become long and thin after some steps of refinement.

The paper is structured as follows. In Section 2 we introduce some notations used for the construction of the volume preserving map. In Section 3 we construct the volume preserving maps between the regular 3D octahedron and the 3D ball $\mathbb{B}^3$. In Section 4 we construct a uniform refinable grid of the regular octahedron followed by implementation issues, and its projection onto $\mathbb{B}^3$. Finally, in Section 5 we construct a multiresolution analysis and piecewise constant wavelet bases of $L^2(\mathbb{B}^3)$.

2. Preliminaries

Consider the ball of radius r centered at the origin O, defined as

$$\mathbb{B}^3 = \left\{ (x,y,z) \in \mathbb{R}^3, x^2 + y^2 + z^2 \leq r^2 \right\}$$

and the regular octahedron $\mathbb{K}$ of the same volume, centered at O and with vertices on the coordinate axes

$$\mathbb{K} = \left\{ (x,y,z) \in \mathbb{R}^3, |x| + |y| + |z| \leq a \right\}.$$

Since the volume of the regular octahedron is $4a^3/3$, we have

$$a = r\sqrt[3]{\pi}. \tag{1}$$

The parametric equations of the ball are

$$\begin{aligned} x &= \rho \cos\theta \sin\varphi, \\ y &= \rho \sin\theta \sin\varphi, \\ z &= \rho \cos\varphi, \end{aligned} \tag{2}$$

where $\varphi \in [0, \pi]$ is the colatitude, $\theta \in [0, 2\pi)$ is the longitude and $\rho \in [0, r]$ is the distance to the origin. A simple calculation shows that the volume element of the ball is

$$dV = \rho^2 \sin\varphi \, d\rho \, d\theta \, d\varphi. \tag{3}$$

The ball and the octahedron can be split into eight congruent parts (see Figure 1), each part being situated in one of the eight octants $I_i^\pm$, $i = 0, 1, 2, 3$,

$$\begin{aligned} I_0^+ &= \{(x,y,z), x \geq 0, y \geq 0, z \geq 0\}, & I_0^- &= \{(x,y,z), x \geq 0, y \geq 0, z \leq 0\}, \\ I_1^+ &= \{(x,y,z), x \leq 0, y \geq 0, z \geq 0\}, & I_1^- &= \{(x,y,z), x \leq 0, y \geq 0, z \leq 0\}, \\ I_2^+ &= \{(x,y,z), x \leq 0, y \leq 0, z \geq 0\}, & I_2^- &= \{(x,y,z), x \leq 0, y \leq 0, z \leq 0\}, \\ I_3^+ &= \{(x,y,z), x \geq 0, y \leq 0, z \geq 0\}, & I_3^- &= \{(x,y,z), x \geq 0, y \leq 0, z \leq 0\}. \end{aligned}$$

Let $\mathbb{B}_i^s$ and $\mathbb{K}_i^s$ be the regions of $\mathbb{B}^3$ and $\mathbb{K}$, situated in I_i^s, respectively.

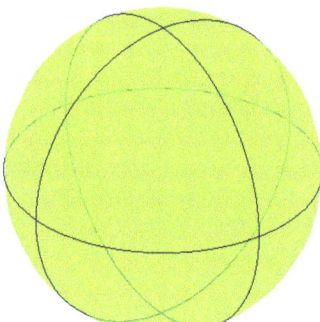

Figure 1. The eight spherical zones obtained as intersections of the coordinate planes with the ball $\mathbb{B}^3$.

Next we will construct a map $\mathcal{U} : \mathbb{B}^3 \to \mathbb{K}$ which preserves the volume, i.e., $\mathcal{U}$ satisfies

$$\text{vol}(D) = \text{vol}(\mathcal{U}(D)), \qquad \text{for all } D \subseteq \mathbb{B}^3, \tag{4}$$

where vol(D) denotes the volume of a domain D. For an arbitrary point $(x, y, z) \in \mathbb{B}^3$ we denote

$$(X, Y, Z) = \mathcal{U}(x, y, z) \in \mathbb{K}. \tag{5}$$

3. Construction of the Volume Preserving Map $\mathcal{U}$ and Its Inverse

We focus on the region $\mathbb{B}_0^+ \subset I_0^+$ where we consider the points $A = (r, 0, 0)$, $B = (0, r, 0)$, $C = (0, 0, r)$ and the vertical plane of equation $y = x \tan \alpha$ with $\alpha \in (0, \pi/2)$ (see Figure 2 (left)). We denote by M its intersection with the great arc $\widehat{AB}$ of the sphere of radius r. More precisely, $M = (r \cos \alpha, r \sin \alpha, 0)$. The volume of the spherical region $OAMC$ equals $r^3 \alpha / 3$.

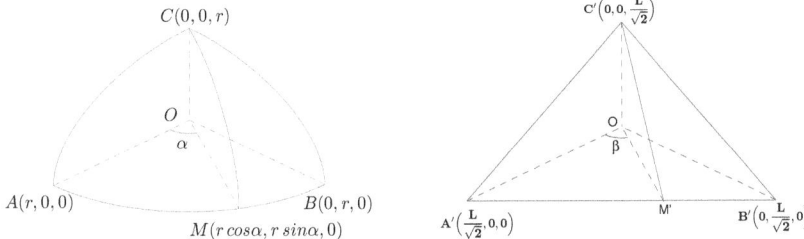

Figure 2. The spherical region $OAMC$ and its image $OA'M'C' = \mathcal{U}(OAMC)$ on the octahedron.

Now we intersect the region $\mathbb{K}_1^+$ of the octahedron with the vertical plane of equation $y = x \tan \beta$ and denote by $M'(m, n, 0)$ its intersection with the edge $A'B'$, where $A'(a, 0, 0)$, $B'(0, a, 0)$ (see Figure 2 (right)). Then $m + n = a$ and from $n = m \tan \beta$ we find

$$m = a \cdot \frac{1}{1 + \tan \beta}, \quad n = a \cdot \frac{\tan \beta}{1 + \tan \beta}.$$

The volume of $OA'M'C'$ is

$$\mathcal{V}(OA'M'C') = \frac{OC' \cdot \mathcal{A}(OA'M')}{3} = \frac{a}{3} \cdot \frac{OA' \cdot n}{2} = \frac{a^3 \tan \beta}{6(1 + \tan \beta)}.$$

If we want the volume of the region $OAMC$ of the unit ball to be equal to the volume of $OA'M'C'$, we obtain

$$\alpha = \frac{\pi}{2} \cdot \frac{\tan \beta}{1 + \tan \beta}, \quad \text{whence } \tan \beta = \frac{2\alpha}{\pi - 2\alpha}.$$

This give us a first relation between (x, y, z) and (X, Y, Z):

$$\frac{Y}{X} = \frac{2 \arctan \frac{y}{x}}{\pi - 2 \arctan \frac{y}{x}}.$$

Using the spherical coordinates (2) we obtain

$$Y = \frac{2\theta}{\pi - 2\theta} \cdot X. \tag{6}$$

In order to obtain a second relation between (x, y, z) and (X, Y, Z), we impose that, for an arbitrary $\widetilde{\rho} \in (0, r]$ the region

$$\left\{ (x, y, z) \in \mathbb{R}^3,\ x^2 + y^2 + z^2 \leq \widetilde{\rho}^2,\ x, y, z \geq 0 \right\} \text{ of volume } \frac{\pi \widetilde{\rho}^3}{6}$$

is mapped by $\mathcal{U}$ onto

$$\{(X,Y,Z) \in \mathbb{R}^3,\ X+Y+Z \leq \ell,\ X,Y,Z \geq 0\} \text{ of volume } \frac{\ell^3}{6}.$$

Then, the volume preserving condition (4) implies $\ell = a \cdot \tilde{\rho}/r$, with a satisfying (1). Thus,

$$X + Y + Z = \frac{a}{r}\sqrt{x^2 + y^2 + z^2}$$

and in spherical coordinates this can be written as

$$X + Y + Z = \frac{a\rho}{r}. \tag{7}$$

In order to have a volume preserving map, the modulus of the Jacobian $J(\mathcal{U})$ of $\mathcal{U}$ must be 1, or, equivalently, taking into account the volume element (3), we must have

$$J(\mathcal{U}) = \begin{vmatrix} X'_\rho & X'_\varphi & X'_\theta \\ Y'_\rho & Y'_\varphi & Y'_\theta \\ Z'_\rho & Z'_\varphi & Z'_\theta \end{vmatrix} = \rho^2 \sin \varphi. \tag{8}$$

Taking into account formula (7), we have

$$J(\mathcal{U}) = \begin{vmatrix} X'_\rho & X'_\varphi & X'_\theta \\ Y'_\rho & Y'_\varphi & Y'_\theta \\ a/r - X'_\rho - Y'_\rho & -X'_\varphi - Y'_\varphi & -X'_\theta - Y'_\theta \end{vmatrix} = \begin{vmatrix} X'_\rho & X'_\varphi & X'_\theta \\ Y'_\rho & Y'_\varphi & Y'_\theta \\ a/r & 0 & 0 \end{vmatrix} = \frac{a}{r} \begin{vmatrix} X'_\varphi & X'_\theta \\ Y'_\varphi & Y'_\theta \end{vmatrix}.$$

Further, using relation (6) we get

$$J(\mathcal{U}) = \frac{a}{r} \begin{vmatrix} X'_\varphi & X'_\theta \\ \frac{2\theta}{\pi-2\theta} \cdot X'_\varphi & \frac{2\theta}{\pi-2\theta} \cdot X'_\theta + \frac{2\pi}{(\pi-2\theta)^2} \cdot X \end{vmatrix} = \frac{a}{r} \begin{vmatrix} X'_\varphi & X'_\theta \\ 0 & \frac{2\pi}{(\pi-2\theta)^2} \cdot X \end{vmatrix} = \frac{2\pi a}{r(\pi-2\theta)^2} X X'_\varphi.$$

For the last equality, we have multiplied the first row by $-2\theta/(\pi - 2\theta)$ and we have added it to the second row. Then, using the expression for $J(\mathcal{U})$ obtained in (8) we get the differential equation

$$2 X'_\varphi \cdot X = \frac{r\rho^2}{\pi a} (\pi - 2\theta)^2 \sin \varphi.$$

The integration with respect to φ gives

$$X^2 = -\frac{r(\pi - 2\theta)^2}{\pi a} \rho^2 \cos \varphi + C(\theta, \rho),$$

and further, for finding $C(\theta, \rho)$ we use the fact that, for $\varphi = \pi/2$ we must obtain $Z = 0$. Thus, for $\varphi = \pi/2$ we have

$$X^2 = C(\theta, \rho), \text{ so } Y = \frac{2\theta}{\pi - 2\theta} \sqrt{C(\theta, \rho)}, \text{ and}$$

$$Z = \frac{a\rho}{r} - X - Y = \frac{a\rho}{r} - \frac{\pi}{\pi - 2\theta} \sqrt{C(\theta, \rho)}.$$

Thus, $Z = 0$ is obtained for

$$C(\theta, \rho) = \frac{a^2 \rho^2}{\pi^2 r^2} (\pi - 2\theta)^2,$$

and finally, the map $\mathcal{U}$ restricted to the region I_0^+ is

$$X = \frac{\sqrt{2}}{\pi^{2/3}} \cdot \rho(\pi - 2\theta) \sin \frac{\varphi}{2}, \qquad (9)$$

$$Y = \frac{\sqrt{2}}{\pi^{2/3}} \cdot \rho \cdot 2\theta \sin \frac{\varphi}{2}, \qquad (10)$$

$$Z = \pi^{1/3} \rho (1 - \sqrt{2} \sin \frac{\varphi}{2}). \qquad (11)$$

In the other seven octants, the map $\mathcal{U}$ can be obtained by symmetry as follows. A point $(x,y,z) \in \mathbb{B}^3$, can be written as

$$(x,y,z) = (\text{sgn}(x) \cdot |x|, \text{sgn}(y) \cdot |y|, \text{sgn}(z) \cdot |z|), \quad \text{with } (|x|, |y|, |z|) \in I_0^+.$$

Therefore, if we denote by $(\overline{X}, \overline{Y}, \overline{Z}) = \mathcal{U}(|x|, |y|, |z|)$, then we can define $\mathcal{U}(x,y,z)$ as

$$\mathcal{U}(x,y,z) = (\text{sgn}(x) \cdot \overline{X}, \text{sgn}(y) \cdot \overline{Y}, \text{sgn}(z) \cdot \overline{Z}). \qquad (12)$$

Next we deduce the formulas for the inverse of $\mathcal{U}$. First, from (6) we obtain

$$\theta = \frac{\pi Y}{2(X+Y)},$$

and from (7) we have

$$\rho = \frac{r}{a}(X+Y+Z) = \pi^{-1/3}(X+Y+Z).$$

Adding (9) and (10), after some more calculations we obtain

$$\sin \frac{\varphi}{2} = \frac{X+Y}{\sqrt{2}(X+Y+Z)},$$

and further

$$\cos \varphi = \frac{Z(2X+2Y+Z)}{(X+Y+Z)^2}, \quad \sin \varphi = \frac{X+Y}{X+Y+Z} \sqrt{2 - \left(\frac{X+Y}{X+Y+Z}\right)^2}.$$

Finally, the inverse $\mathcal{U}^{-1} : \mathbb{K} \to \mathbb{B}^3$ is defined by

$$x = \pi^{-1/3}(X+Y)\sqrt{2 - \left(\frac{X+Y}{X+Y+Z}\right)^2} \cos \frac{\pi Y}{2(X+Y)}, \qquad (13)$$

$$y = \pi^{-1/3}(X+Y)\sqrt{2 - \left(\frac{X+Y}{X+Y+Z}\right)^2} \sin \frac{\pi Y}{2(X+Y)}, \qquad (14)$$

$$z = \pi^{-1/3} \frac{Z(2X+2Y+Z)}{(X+Y+Z)}. \qquad (15)$$

for $(X,Y,Z) \in \mathbb{K}_0^+$, and for the other seven octants the formulas can be calculated as in (12).

4. Uniform and Refinable Grids of the Regular Octahedron and of the Ball

In this section we construct a uniform refinement of the regular octahedron $\mathbb{K}$ of volume $\text{vol}(\mathbb{K})$, more precisely a subdivision of $\mathbb{K}$ into 64 cells of two shapes, each of them having the volume $\text{vol}(\mathbb{K})/64$. This subdivision can be repeated for each of the 64 small cells, the resulting 64^2 cells of volume $\text{vol}(\mathbb{K})/64^2$ being of one of the two types from the first refinement. Next, the volume preserving map $\mathcal{U}$ will allow us the construction of uniform and refinable grids of the 3D ball $\mathbb{B}^3$ by

Mathematics **2020**, *8*, 994

transporting the octahedral uniform refinable 3D grids, and further, the construction of orthonormal piecewise constant wavelets on the 3D ball.

4.1. Refinement of the Octahedron

The initial octahedron $\mathbb{K}$ consists in four congruent cells, each situated in one of the octants $I_i^+ \cup I_i^-$, $i = 0, 1, 2, 3$ (see Figure 3). We will say that this type of cell is $\mathbf{T}_0$, the index 0 of $\mathbf{T}_0$ being the coarsest level of the refinement. For simplifying the writing we denote by $\mathbb{N}_0$ the set of positive natural numbers and by $\mathbb{N}_n = \{1, 2, \ldots, n\}$, for $n \in \mathbb{N}_0$.

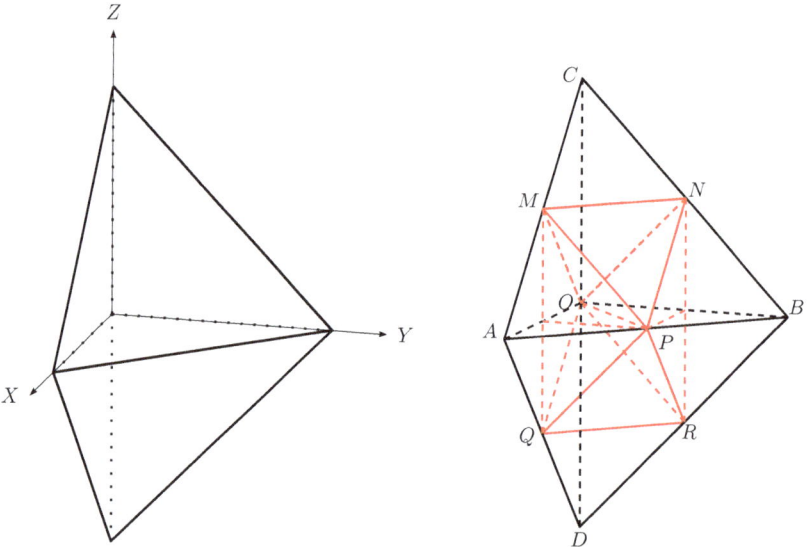

Figure 3. Left: one of the four cells of type $\mathbf{T}_0$ constituting the octahedron. Right: each cell of type $\mathbf{T}_0$ can be subdivided into six cells of type $\mathbf{T}_1$ and two cells of type $\mathbf{M}_1$.

4.1.1. First Step of Refinement

The cell $\mathbf{T}_0 = (ABCD) \in I_0^+ \cup I_0^-$, with $A(a, 0, 0)$, $B(0, a, 0)$, $C(0, 0, a)$, $D(0, 0, -a)$ (see Figure 3), will be subdivided into eight smaller cells having the same volume, as follows: we take the mid-points M, N, P, Q, R of the edges AC, BC, AB, AD, BD, respectively. Thus, one obtains $t_1 = 6$ cells of type $\mathbf{T}_1$ (MQOP, MQAP, NROP, NRBP, ODQR and COMN), and $m_1 = 2$ other cells, OMNP and OPQR, of another type, say $\mathbf{M}_1$. The cells of type $\mathbf{T}_1$ have the same shape with the cells $\mathbf{T}_0$. Their volumes are

$$\text{vol}(\mathbf{T}_1) = \text{vol}(\mathbf{M}_1) = \frac{\text{vol}(\mathbf{T}_0)}{8}.$$

Figures 4 and 5 also show the eight cells at the first step of refinement.

Similarly we refine the other three cells situated in $I_i^+ \cup I_i^-$, $i = 1, 2, 3$, therefore the total number of cells after the first step of refinement is 32, more precisely 24 of type $\mathbf{T}_1$ and 8 of type $\mathbf{M}_1$.

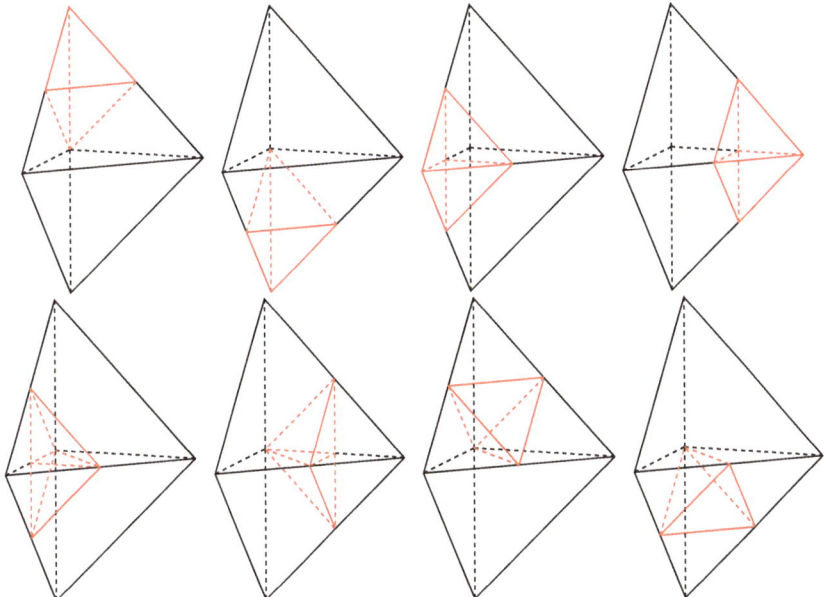

Figure 4. The subdivision of a **T** cell.

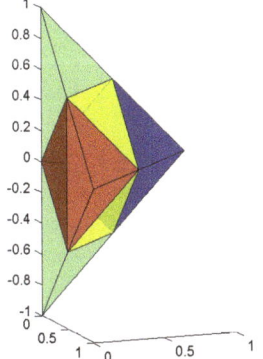

Figure 5. The first step of the refinement: the cell T_0 is divided into two cells of type M_1 (yellow) and six cells of type T_1: two red, two blue and two green.

4.1.2. Second Step of Refinement

A cell of type T_1 will be subdivided in the same way as a cell of type T_0, i.e., into six cells of type T_2 and two cells of type M_2. Their volumes will be

$$\text{vol}(T_2) = \text{vol}(M_2) = \frac{\text{vol}(T_0)}{8^2}.$$

Therefore, from the subdivision of the 6 cells of type T_1 we have 36 cells of type T_2 and 12 cells of type M_2.

For a cell $(OMNP)$ of type M_1, which is a regular tetrahedron of edge $\ell_1 = a\sqrt{2}/2$, we take the mid-points of the six edges (see Figures 6 and 7). This will give four cells of type T_2 in the middle and four cells of type M_2, i.e., regular tetrahedrons of edge $\ell_2 = a\sqrt{2}/2^2$. From the subdivision of the two cells of type M_1 we have 8 cells of type T_2 and 8 cells of type M_2.

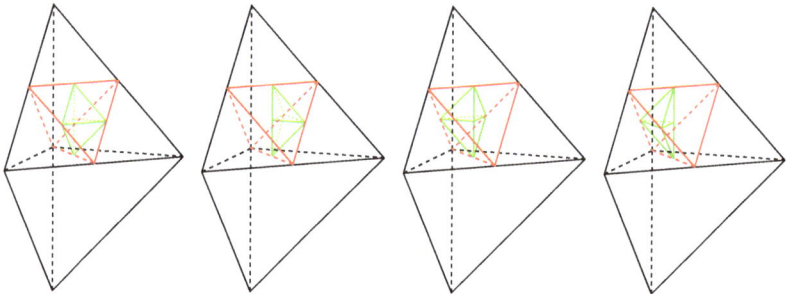

Figure 6. The four cells of type **T** of the subdivision of a cell of type **M**.

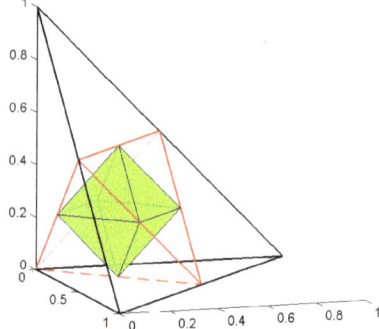

Figure 7. The subdivision of a cell of type M_1 into four cells of type **M**: the four tetrahedrons at the corners and four cells of type **T** in the middle, forming an octahedron.

In conclusion, the second step of subdivision yields in $I_0^+ \cup I_0^- \ t_2 = 44$ cells of type T_2 and $m_2 = 20$ cells of type M_2, each having the volume $\text{vol}(T_0)/64$, therefore the total number of cells after the second refinement will be $4 \cdot 8^2$, more precisely 76 of type T_2 and 80 of type M_2.

4.1.3. The General Step of Refinement

Let m_j and t_j denote the numbers of cells of type M_j and T_j, respectively, resulted at the step j of the subdivision, starting from one cell of type T_0. At this step, each of the t_{j-1} cells of type T_{j-1} is subdivided into 6 cells of type T_j and 2 cells of type M_j, and each of the m_{j-1} cells of type M_{j-1} is subdivided into 4 cells of type T_j and 4 cells of type M_j. This implies

$$t_j = 6t_{j-1} + 4m_{j-1},$$
$$m_j = 2t_{j-1} + 4m_{j-1},$$

or

$$\begin{pmatrix} t_j \\ m_j \end{pmatrix} = A \begin{pmatrix} t_{j-1} \\ m_{j-1} \end{pmatrix} = A^2 \begin{pmatrix} t_{j-2} \\ m_{j-2} \end{pmatrix} = \ldots = A^j \begin{pmatrix} t_0 \\ m_0 \end{pmatrix},$$

with $t_0 = 1$, $m_0 = 0$ and $A = \begin{pmatrix} 6 & 4 \\ 2 & 4 \end{pmatrix}$. After some calculations we obtain

$$A^j = \frac{1}{3} \begin{pmatrix} 2^j(2^{2j+1}+1) & 2^{j+1}(2^{2j}-1) \\ 2^j(2^{2j}-1) & 2^j(2^{2j}+2) \end{pmatrix}, \text{ whence}$$

$$t_j = \frac{2^j}{3}(2^{2j+1}+1), \quad m_j = \frac{2^j}{3}(2^{2j}-1),$$

the total number of cells of $\mathbb{K}_1^+ \cup \mathbb{K}_1^-$ at step j being $t_j + m_j = 8^j$, and $4 \cdot 8^j$ for the whole octahedron $\mathbb{K}$. Each of the cells of type $\mathbf{T}_j$ and $\mathbf{M}_j$ has the volume $\text{vol}(\mathbf{T}_0)/8^j$.

4.2. Implementation Issues

Every cell of the polyhedron is identified by the coordinates of its four vertices. We have two types of cells, which will be denoted by **T** and **M**.

A cell of type **T** has the same coordinates x and y for the first two vertices. The z coordinate of the first vertex is greater than the z coordinate of the second vertex and the mean value of these z coordinates gives the value of the z coordinate of the third and fourth vertices of **T**.

A cell of type **M** has two pairs of vertices at the same altitude (the same value of the z coordinate).

At every step of refinement, every cell **T** is divided into 6 cells of type **T** and two cells of type **M**. Suppose $[\mathbf{p}_1, \mathbf{p}_2, \mathbf{p}_3, \mathbf{p}_4]$ is the array giving the coordinates of the four vertices of a T cell. The coordinates of the vertices of the next level cells are computed as follows

next level cell number 1 : $\frac{1}{2}[\mathbf{p}_1 + \mathbf{p}_1, \mathbf{p}_1 + \mathbf{p}_2, \mathbf{p}_1 + \mathbf{p}_3, \mathbf{p}_1 + \mathbf{p}_4]$,

next level cell number 2 : $\frac{1}{2}[\mathbf{p}_2 + \mathbf{p}_1, \mathbf{p}_2 + \mathbf{p}_2, \mathbf{p}_2 + \mathbf{p}_3, \mathbf{p}_2 + \mathbf{p}_4]$,

next level cell number 3 : $\frac{1}{2}[\mathbf{p}_3 + \mathbf{p}_1, \mathbf{p}_3 + \mathbf{p}_2, \mathbf{p}_3 + \mathbf{p}_3, \mathbf{p}_3 + \mathbf{p}_4]$,

next level cell number 4 : $\frac{1}{2}[\mathbf{p}_4 + \mathbf{p}_1, \mathbf{p}_4 + \mathbf{p}_2, \mathbf{p}_4 + \mathbf{p}_3, \mathbf{p}_4 + \mathbf{p}_4]$,

next level cell number 5 : $\frac{1}{2}[\mathbf{p}_1 + \mathbf{p}_3, \mathbf{p}_2 + \mathbf{p}_3, \mathbf{p}_3 + \mathbf{p}_3, \mathbf{p}_4 + \mathbf{p}_1 + \mathbf{p}_2]$,

next level cell number 6 : $\frac{1}{2}[\mathbf{p}_1 + \mathbf{p}_4, \mathbf{p}_2 + \mathbf{p}_4, \mathbf{p}_4 + \mathbf{p}_1 + \mathbf{p}_2, \mathbf{p}_3 + \mathbf{p}_4]$,

next level cell number 7 : $\frac{1}{2}[\mathbf{p}_1 + \mathbf{p}_2, \mathbf{p}_1 + \mathbf{p}_3, \mathbf{p}_1 + \mathbf{p}_4, \mathbf{p}_3 + \mathbf{p}_4]$,

next level cell number 8 : $\frac{1}{2}[\mathbf{p}_1 + \mathbf{p}_2, \mathbf{p}_2 + \mathbf{p}_3, \mathbf{p}_2 + \mathbf{p}_4, \mathbf{p}_3 + \mathbf{p}_4]$.

The cells 1–6 are of type **T** and the cells 7 and 8 are of type **M** (see Figure 4).

Every cell **M** consists in 4 cells of type **T** and 4 cells of type **M**. Suppose $[\mathbf{p}_1, \mathbf{p}_2, \mathbf{p}_3, \mathbf{p}_4]$ is the array giving the coordinates of the four vertices of the cell **M** and let $\mathbf{p}_k = (p_{kx}, p_{ky}, p_{kz})$, $k = 1, 2, 3, 4$. We rearrange these four vertices in ascending order with respect to the z coordinate. Let $[\mathbf{q}_1, \mathbf{q}_2, \mathbf{q}_3, \mathbf{q}_4]$ be the vector $[\mathbf{p}_1, \mathbf{p}_2, \mathbf{p}_3, \mathbf{p}_4]$ sorted ascendingly with respect to the z coordinate of the vertices, i.e., $q_{1z} \leq q_{2z} \leq q_{3z} \leq q_{4z}$. Similarly, let $[\mathbf{r}_1, \mathbf{r}_2, \mathbf{r}_3, \mathbf{r}_4]$ be the rearrangement of vertices $\mathbf{p}_1, \ldots, \mathbf{p}_4$ such that $r_{1x} \leq r_{2x} \leq r_{3x} \leq r_{4x}$. Let, also, $[\mathbf{s}_1, \mathbf{s}_2, \mathbf{s}_3, \mathbf{s}_4]$ be the array of rearranged vertices with respect to the y

coordinate in ascending order. The coordinates of the vertices of the cells at the next level are computed as follows:

$$\text{next level cell number 1}: \quad \frac{1}{2}[q_3+q_4, q_1+q_2, r_3+r_4, s_3+s_4]$$

$$\text{next level cell number 2}: \quad \frac{1}{2}[q_3+q_4, q_1+q_2, s_3+s_4, r_1+r_2]$$

$$\text{next level cell number 3}: \quad \frac{1}{2}[q_3+q_4, q_1+q_2, r_1+r_2, s_1+s_2]$$

$$\text{next level cell number 4}: \quad \frac{1}{2}[q_3+q_4, q_1+q_2, s_1+s_2, r_3+r_4]$$

$$\text{next level cell number 5}: \quad \frac{1}{2}[\mathbf{p}_1+\mathbf{p}_1, \mathbf{p}_1+\mathbf{p}_2, \mathbf{p}_1+\mathbf{p}_3, \mathbf{p}_1+\mathbf{p}_4]$$

$$\text{next level cell number 6}: \quad \frac{1}{2}[\mathbf{p}_2+\mathbf{p}_1, \mathbf{p}_2+\mathbf{p}_2, \mathbf{p}_2+\mathbf{p}_3, \mathbf{p}_2+\mathbf{p}_4]$$

$$\text{next level cell number 7}: \quad \frac{1}{2}[\mathbf{p}_3+\mathbf{p}_1, \mathbf{p}_3+\mathbf{p}_2, \mathbf{p}_3+\mathbf{p}_3, \mathbf{p}_3+\mathbf{p}_4]$$

$$\text{next level cell number 8}: \quad \frac{1}{2}[\mathbf{p}_4+\mathbf{p}_1, \mathbf{p}_4+\mathbf{p}_2, \mathbf{p}_4+\mathbf{p}_3, \mathbf{p}_4+\mathbf{p}_4].$$

To verify whether a point $\mathbf{p}=(p_x, p_y, p_z)$ is inside a cell with vertices $[\mathbf{p}_1, \mathbf{p}_2, \mathbf{p}_3, \mathbf{p}_4]$, we compute the following numbers:

$$d_1 = \text{sgn} \begin{vmatrix} p_{1x} & p_{2x} & p_{3x} & p_x \\ p_{1y} & p_{2y} & p_{3y} & p_y \\ p_{1z} & p_{2z} & p_{3z} & p_z \\ 1 & 1 & 1 & 1 \end{vmatrix}, d_2 = \text{sgn} \begin{vmatrix} p_{1x} & p_{2x} & p_x & p_{4x} \\ p_{1y} & p_{2y} & p_y & p_{4y} \\ p_{1z} & p_{2z} & p_z & p_{4z} \\ 1 & 1 & 1 & 1 \end{vmatrix}, d_3 = \text{sgn} \begin{vmatrix} p_{1x} & p_x & p_{3x} & p_{4x} \\ p_{1y} & p_y & p_{3y} & p_{4y} \\ p_{1z} & p_z & p_{3z} & p_{4z} \\ 1 & 1 & 1 & 1 \end{vmatrix},$$

$$d_4 = \text{sgn} \begin{vmatrix} p_x & p_{2x} & p_{3x} & p_{4x} \\ p_y & p_{2y} & p_{3y} & p_{4y} \\ p_z & p_{2z} & p_{3z} & p_{4z} \\ 1 & 1 & 1 & 1 \end{vmatrix}, d_5 = \text{sgn} \begin{vmatrix} p_{1x} & p_{2x} & p_{3x} & p_{4x} \\ p_{1y} & p_{2y} & p_{3y} & p_{4y} \\ p_{1z} & p_{2z} & p_{3z} & p_{4z} \\ 1 & 1 & 1 & 1 \end{vmatrix}.$$

We calculate $v = |d_1| + |d_2| + |d_2| + |d_3| + |d_4| + |d_5|$. If $|d_1 + d_2 + d_3 + d_4 + d_5| = v$, then for $v=5$ the point $\mathbf{p}$ is in the interior of the cell, for $v=4$ the point $\mathbf{p}$ is on one of the faces of the cell, for $v=3$ the point $\mathbf{p}$ is situated on one of the edges of the cell, and for $v=2$ the point $\mathbf{p}$ is one of the vertices of the cell. If $|d_1 + d_2 + d_3 + d_4 + d_5| \neq v$, the point $\mathbf{p}$ is located outside the cell. Since the vertices $\mathbf{p}_k$ are different we have $v \geq 2$.

4.3. Uniform and Refinable Grids of the Ball $\mathbb{B}^3$

If we transport the uniform and refinable grid on $\mathbb{K}$ onto the ball $\mathbb{B}^3$ using the volume preserving map $\mathcal{U}^{-1}$, we obtain a uniform and refinable grid of $\mathbb{B}^3$. Figures 8–10 show the images on $\mathbb{B}^3$ of different cells of $\mathbb{K}$.

Besides the multiresolution analysis and wavelet bases, which will be constructed in Section 5, another useful application is the construction of a uniform sampling of the rotation group $SO(3)$, by calculations similar to the ones in [3]. This will be subject of a future paper.

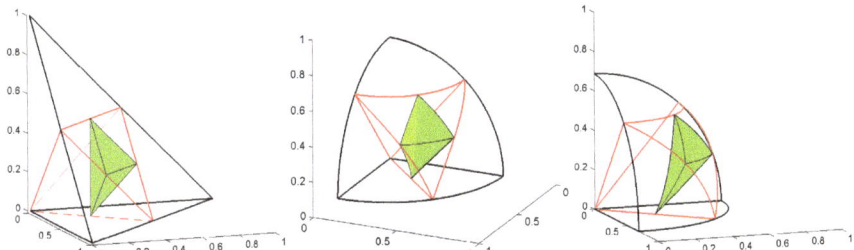

Figure 8. Left: a cell of **M** in red and a cell of **T** type in green from the octahedron Middle and right: the corresponding cells of the ball.

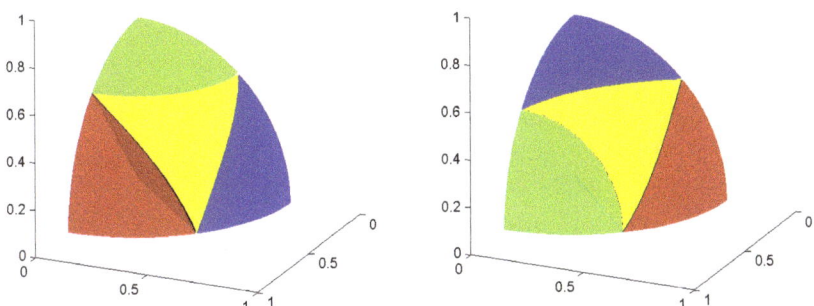

Figure 9. Left: the image on the ball of the positive octant; Right: the same image rotated.

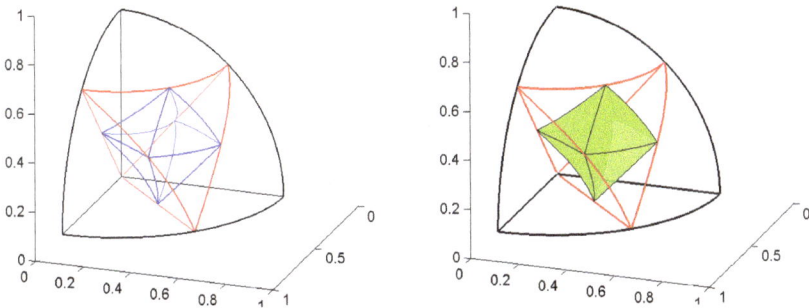

Figure 10. The image on the ball of the cells of the octahedron corresponding to Figure 7.

5. Multiresolution Analysis and Piecewise Constant Orthonormal Wavelet Bases of $L^2(\mathbb{K})$ and $L^2(\mathbb{B}^3)$

Let $\mathcal{D} = \mathcal{D}^0 = \{D_1, D_2, D_3, D_4\}$ be the decomposition of the domain $\mathbb{K}$ considered in Section 4.1, consisting in four congruent domains (cells) of type $\mathbf{T}_0$. For $D \in \mathcal{D}$, let $\mathcal{R}_D$ denote the set of the eight refined domains, constructed in Section 4.1.1. The set $\mathcal{D}^1 = \cup_{D \in \mathcal{D}^0} \mathcal{R}_D$ is a refinement of $\mathcal{D}^0$, consisting in $4 \cdot 8$ congruent cells. Continuing the refinement process as we described in Section 4, we obtain a decomposition $\mathcal{D}^j$ of $\mathbb{K}$, for $j \in \mathbb{N}_0$, $|\mathcal{D}^j| = 4 \cdot 8^j$.

For a fixed $j \in \mathbb{N}_0$ we assign to each domain $D_k^j \in \mathcal{D}^j$, $k \in \mathcal{N}_j := \mathbb{N}_{4 \cdot 8^j}$, the function $\varphi_{D_k^j} : \mathbb{K} \to \mathbb{R}$,

$$\varphi_{D_k^j} = (2\sqrt{2})^j \frac{2}{\sqrt{\text{vol}(\mathbb{K})}} \chi_{D_k^j},$$

where $\chi_{D_k^j}$ is the characteristic function of the domain D_k^j. Then we define the spaces of functions $V^j = \text{span}\{\varphi_{D_k^j}, k \in \mathcal{N}_j\}$ of dimension $4 \cdot 8^j$, consisting of piecewise constant functions on the domains of $\mathcal{D}^j$. Moreover, we have $\|\varphi_{D_k^j}\|_{L^2(\mathbb{K})} = 1$, the norm being the usual 2-norm of the space $L^2(\mathbb{K})$. For $A^j \in \mathcal{D}^j = \{D_k^j, j \in \mathcal{N}_j\}$, let A_k^{j+1}, $k \in \mathbb{N}_8$, be the refined subdomains obtained from A^j. One has

$$\varphi_{A^j} = \frac{1}{2\sqrt{2}}\left(\varphi_{A_1^{j+1}} + \varphi_{A_2^{j+1}} + \ldots + \varphi_{A_8^{j+1}}\right),$$

in $L^2(\mathbb{K})$, equality which implies the inclusion $V^j \subseteq V^{j+1}$, for all $j \in \mathbb{N}_0$. With respect to the usual inner product $\langle \cdot, \cdot \rangle_{L^2(\mathbb{K})}$, the spaces V^j are Hilbert spaces, with the corresponding usual 2-norm $\|\cdot\|_{L^2(\mathbb{K})}$. In conclusion, the sequence of subspaces V^j has the following properties:

1. $V^j \subseteq V^{j+1}$ for all $j \in \mathbb{N}_0$,
2. $\text{clos}_{L^2(\mathbb{K})} \bigcup_{j=0}^{\infty} V^j = L^2(\mathbb{K})$,
3. The set $\{\varphi_{D_k^j}, k \in \mathcal{N}_j\}$ is an orthonormal basis of the space V^j for each $j \in \mathbb{N}_0$,

i.e., the sequence $\{V^j, j \in \mathbb{N}_0\}$ constitutes a *multiresolution analysis* of the space $L^2(\mathbb{K})$. Let W^j denote the orthogonal complement of the coarse space V^j in the fine space V^{j+1}, so that

$$V^{j+1} = V^j \oplus W^j.$$

The dimension of W^j is $\dim W^j = 28 \cdot 8^j$. The spaces W^j are called *wavelet spaces* and their elements are called *wavelets*. In the following we construct an orthonormal basis of W^j. To each domain $A^j \in \mathcal{D}^j$, seven wavelets supported on D^j will be associated in the following way:

$$\psi_{A^j}^{\ell} = a_{\ell 1}\varphi_{A_1^{j+1}} + a_{\ell 2}\varphi_{A_2^{j+1}} + \ldots + a_{\ell 8}\varphi_{A_8^{j+1}}, \text{ for } \ell \in \mathbb{N}_7,$$

with $a_{\ell j} \in \mathbb{R}$, $\ell \in \mathbb{N}_7$, $j \in \mathbb{N}_8$. We have to find conditions on the coefficients $a_{\ell j}$ which ensure that the set $\{\psi_{A^j}^{\ell}, \ell \in \mathbb{N}_7, A^j \in \mathcal{D}^j\}$ is an orthonormal basis of W^j. First we must have

$$\langle \psi_{A^j}^{\ell}, \varphi_{S^j} \rangle = 0, \text{ for } \ell \in \mathbb{N}_7 \text{ and } A^j, S^j \in \mathcal{D}^j. \tag{16}$$

If $A^j \neq S^j$, the equality is immediate, since $\text{supp } \psi_{A^j}^{\ell} \subseteq \text{supp } \varphi_{A^j}$ and $\text{supp } \varphi_{A^j} \cap \text{supp } \varphi_{S^j}$ is either empty or an edge, whose measure is zero. If $A^j = S^j$, evaluating the inner product (16) we obtain

$$\langle \psi_{A^j}^{\ell}, \varphi_{S^j} \rangle = \langle a_{\ell 1}\varphi_{A_1^{j+1}} + a_{\ell 2}\varphi_{A_2^{j+1}} + \ldots + a_{\ell 8}\varphi_{A_8^{j+1}}, \varphi_{A^j} \rangle$$

$$= \frac{1}{2\sqrt{2}}(a_{\ell 1} + a_{\ell 2} + \ldots + a_{\ell 8}).$$

Then, each of the orthogonality conditions

$$\langle \psi_{A^j}^{\ell}, \psi_{A^j}^{\ell'} \rangle = \delta_{\ell \ell'}, \text{ for all } A^j \in \mathcal{D}^j,$$

is equivalent to $a_{\ell'1}a_{\ell 1} + a_{\ell'2}a_{\ell 2} + \ldots + a_{\ell'8}a_{\ell 8} = \delta_{\ell \ell'}$, $\ell, \ell' \in \mathbb{N}_7$. In fact, one requires the orthogonality of the 8×8 matrix $M = (a_{ij})_{i,j}$ with the entries of the first row equal to $1/(2\sqrt{2})$.

A particular case was considered in [12], where the authors divide a tetrahedron into eight small tetrahedrons of the same area using Bey's method and for the construction of the orthonormal wavelet basis they take the Haar matrix

$$\frac{1}{2\sqrt{2}} \begin{pmatrix} 1 & 1 & 1 & 1 & 1 & 1 & 1 & 1 \\ 1 & 1 & 1 & 1 & -1 & -1 & -1 & -1 \\ 1 & 1 & -1 & -1 & 0 & 0 & 0 & 0 \\ 0 & 0 & 0 & 0 & 1 & 1 & -1 & -1 \\ 1 & -1 & 0 & 0 & 0 & 0 & 0 & 0 \\ 0 & 0 & 1 & -1 & 0 & 0 & 0 & 0 \\ 0 & 0 & 0 & 0 & 1 & -1 & 0 & 0 \\ 0 & 0 & 0 & 0 & 0 & 0 & 1 & -1 \end{pmatrix}$$

Alternatively, we can consider the symmetric orthogonal matrix

$$\begin{pmatrix} c & c & c & c & c & c & c & c \\ c & a & b & b & b & b & b & b \\ c & b & a & b & b & b & b & b \\ c & b & b & a & b & b & b & b \\ c & b & b & b & a & b & b & b \\ c & b & b & b & b & a & b & b \\ c & b & b & b & b & b & a & b \\ c & b & b & b & b & b & b & a \end{pmatrix},$$

with

$$a = \frac{\pm 24 - \sqrt{2}}{28}, \quad b = \frac{\mp 4 - \sqrt{2}}{28}, \quad c = \frac{1}{2\sqrt{2}},$$

or the tensor product $H \otimes H \otimes H$ of the matrix

$$H = \frac{1}{\sqrt{2}} \begin{pmatrix} 1 & 1 \\ 1 & -1 \end{pmatrix}, \text{ which is}$$

$$\frac{1}{2\sqrt{2}} \begin{pmatrix} 1 & 1 & 1 & 1 & 1 & 1 & 1 & 1 \\ 1 & -1 & 1 & -1 & 1 & -1 & 1 & -1 \\ 1 & 1 & -1 & -1 & 1 & 1 & -1 & -1 \\ 1 & -1 & -1 & 1 & 1 & -1 & -1 & 1 \\ 1 & 1 & 1 & 1 & -1 & -1 & -1 & -1 \\ 1 & -1 & 1 & -1 & -1 & 1 & -1 & 1 \\ 1 & 1 & -1 & -1 & -1 & -1 & 1 & 1 \\ 1 & -1 & -1 & 1 & -1 & 1 & 1 & -1 \end{pmatrix}$$

or, more general, we can generate *all* orthogonal 8×8 matrices with the entries of the first row equal to $1/(2\sqrt{2})$ using the method described in [13], where we start with the well known Euler's formula for the general form of a 3×3 rotation matrix. It is also possible to use different orthogonal matrices for the wavelets associated to the decomposition of the cells of type **T** and **M**.

Next, following the ideas in [14] we show how one can transport the above multiresolution analysis and wavelet bases on the 3D ball $\mathbb{B}^3$, using the volume preserving map $\mathcal{U}: \mathbb{B}^3 \to \mathbb{K}$ constructed in Section 3.

Consider the ball $\mathbb{B}^3$ is given by the parametric equations

$$\zeta = \zeta(X, Y, Z) = \mathcal{U}^{-1}(X, Y, Z) = (x(X, Y, Z), y(X, Y, Z), z(X, Y, Z)),$$

with $(X, Y, Z) \in \mathbb{K}$. Since $\mathcal{U}$ and its inverse preserve the volume, the volume element $d\omega(\xi)$ of $\mathbb{B}^3$ equals the volume element $dX\,dY\,dZ = d\mathbf{x}$ of $\mathbb{K}$ (and $\mathbb{R}^3$). Therefore, for all $\widetilde{f}, \widetilde{g} \in L^2(\mathbb{B}^3)$ we have

$$\begin{aligned}
\langle \widetilde{f}, \widetilde{g} \rangle_{L^2(\mathbb{B}^3)} &= \int_{\mathbb{B}^3} \overline{\widetilde{f}(\xi)} \widetilde{g}(\xi)\, d\omega(\xi) \\
&= \int_{\mathcal{U}(\mathbb{B}^3)} \overline{\widetilde{f}(\mathcal{U}^{-1}(X,Y,Z))}\, \widetilde{g}(\mathcal{U}^{-1}(X,Y,Z))\, dX\,dY\,dZ \\
&= \langle \widetilde{f} \circ \mathcal{U}^{-1}, \widetilde{g} \circ \mathcal{U}^{-1} \rangle_{L^2(\mathbb{K})},
\end{aligned}$$

and similarly, for all $f, g \in L^2(\mathbb{K})$ we have

$$\langle f, g \rangle_{L^2(\mathbb{K})} = \langle f \circ \mathcal{U}, g \circ \mathcal{U} \rangle_{L^2(\mathbb{B}^3)}. \tag{17}$$

If we consider the map $\Pi : L^2(\mathbb{B}^3) \to L^2(\mathbb{K})$ induced by $\mathcal{U}$, defined by

$$(\Pi \widetilde{f})(X,Y,Z) = \widetilde{f}\left(\mathcal{U}^{-1}(X,Y,Z)\right), \text{ for all } \widetilde{f} \in L^2(\mathbb{B}^3),$$

and its inverse $\Pi^{-1} : L^2(\mathbb{K}) \to L^2(\mathbb{B}^3)$,

$$(\Pi^{-1} f)(\xi) = f(\mathcal{U}(\xi)), \text{ for all } f \in L^2(\mathbb{K}),$$

then Π is a unitary map, that is

$$\langle \Pi \widetilde{f}, \Pi \widetilde{g} \rangle_{L^2(\mathbb{K})} = \langle \widetilde{f}, \widetilde{g} \rangle_{L^2(\mathbb{B}^3)}, \tag{18}$$

$$\langle \Pi^{-1} f, \Pi^{-1} g \rangle_{L^2(\mathbb{B}^3)} = \langle f, g \rangle_{L^2(\mathbb{K})}. \tag{19}$$

Equality (17) suggests us the construction of orthonormal scaling functions and wavelets defined on $\mathbb{B}^3$. The scaling functions $\widetilde{\varphi}_{D^j_k} : \mathbb{B}^3 \to \mathbb{R}$ will be

$$\widetilde{\varphi}_{D^j_k} = \varphi_{D^j_k} \circ \mathcal{U} = \begin{cases} 1, & \text{on } \mathcal{U}^{-1}(D^j_k), \\ 0, & \text{in rest.} \end{cases} \tag{20}$$

and the wavelets will be defined similarly,

$$\widetilde{\psi}^{\ell}_{A^j} = \psi^{\ell}_{A^j} \circ \mathcal{U}.$$

From equality (17) we can conclude that the spaces

$$\widetilde{V^j} := \operatorname{span}\{\widetilde{\varphi}_{D^j_k}, k \in \mathcal{N}_j\}$$

constitute a multiresolution analysis of $L^2(\mathbb{B}^3)$, each of the set $\{\widetilde{\varphi}_{D^j_k}, k \in \mathcal{N}_j\}$ being an orthonormal basis for the space $\widetilde{V^j}$. Moreover, the set

$$\{\widetilde{\psi}^{\ell}_{A^j}, \ell \in \mathbb{N}_7, A^j \in \mathcal{D}_j\}$$

is an orthonormal basis of $\widetilde{W^j}$.

6. Conclusions and Future Works

The 3D uniform hierarchical grid constructed here can find applications in texture analysis of crystalls, by constructing a grid in the space of 3D rotations, using the technique used in [3]. A comparison of these grids is subject of a future paper.

Another interesting topic which we are going to approach in the future is to compare our wavelets with other 3D wavelets on the ball, listed in the introduction.

Author Contributions: Conceptualization, writing, visualization, A.H. and D.R. All authors have read and agreed to the published version of the manuscript.

Funding: This research received no external funding.

Conflicts of Interest: The authors declare no conflict of interest.

References

1. Moons, T.; van Gool, L.; Vergauwen, M. 3D reconstruction from multiple images part 1: Principles. *Found. Trends Comput. Graph. Vis.* **2010**, *4*, 287–404. [CrossRef]
2. Kendrew, J.C.; Bodo, G.; Dintzis, H.M.; Parrish, R.G.; Wyckoff, H.; Phillips, D.C. A Three-Dimensional Model of the Myoglobin Molecule Obtained by X-Ray Analysis. *Nature* **1958**, *181*, 662–666. [CrossRef] [PubMed]
3. Roşca, D.; Morawiec, A.; de Graef, M. A new method of constructing a grid in the space of 3D rotations and its applications to texture analysis. *Model. Simul. Mater. Sci. Eng.* **2014**, *22*, 075013. [CrossRef]
4. Flenger, M.; Michel, D.; Michel, V. Harmonic spline-wavelets on the 3D ball and their application to the reconstruction of the Earth's density distribution from gravitational data and arbitrary shaped satellite orbits. *ZAMM J. Appl. Math. Mech.* **2006**, *86*, 856–873.
5. Simons, F.J.; Loris, I.; Nolet, G.; Daubechies, I.; Voronin, S.; Judd, J.S.; Vetter, P.A.; Charléty, J.; Vonesch, C. Solving or resolving global tomographic models with spherical wavelets, and the scale and sparsity of seismic heterogeneity. *Geophys. J. Int.* **2011**, *187*, 969–988. [CrossRef]
6. Simons, F.J.; Loris, I.; Brevdo, E.; Daubechies, I. Wavelets and wavelet-like transforms on the sphere and their application to geophysical data inversion. In *Wavelets and Sparsity XIV*; Papadakis, M., Van de Ville, D., Goyal, V.K., Eds.; International Society for Optics and Photonics: San Diego, CA, USA, 2011; Volume 81380, pp. 1–15.
7. Chow, A. Orthogonal and Symmetric Haar Wavelets on the Three-Dimensional Ball. Master's Thesis, University of Toronto, Toronto, ON, Canada, 2010.
8. Leistedt, B.; McEwen, J.D. Exact wavelets on the ball. *IEEE Trans. Signal Process.* **2012**, *60*, 6257–6269. [CrossRef]
9. Loris, I.; Simons, F.J.; Daubechies, I.; Nolet, G.; Fornasier, M.; Vetter, P.; Judd, S.; Voronin, S.; Vonesch, C.; Charléty, J. A new approach to global seismic tomography based on regularization by sparsity in a novel 3D spherical wavelet basis. In Proceedings of the EGU General Assembly Conference Abstracts, ser. EGU General Assembly Conference Abstracts, Vienna, Austria, 2–7 May 2010; Volume 12, p. 6033.
10. Michel, V. Wavelets on the 3 dimensional ball. *Proc. Appl. Math. Mech.* **2005**, *5*, 775–776. [CrossRef]
11. Roşca, D. Locally supported rational spline wavelets on the sphere. *Math. Comput.* **2005**, *74*, 1803–1829. [CrossRef]
12. Boscardin, L.B.; Castro, L.R.; Castro, S.M. Haar-Like Wavelets over Tetrahedra. *J. Comput. Sci. Technol.* **2017**, *17*, 92–99. [CrossRef]
13. Pop, V.; Roşca, D. Generalized piecewise constant orthogonal wavelet bases on 2D-domains. *Appl. Anal.* **2011**, *90*, 715–723. [CrossRef]
14. Roşca, D. Wavelet analysis on some surfaces of revolution via area preserving projection. *Appl. Comput. Harm. Anal.* **2011**, *30*, 272–282. [CrossRef]

© 2020 by the authors. Licensee MDPI, Basel, Switzerland. This article is an open access article distributed under the terms and conditions of the Creative Commons Attribution (CC BY) license (http://creativecommons.org/licenses/by/4.0/).

Article
A Simple Method for Network Visualization

Jintae Park, Sungha Yoon, Chaeyoung Lee and Junseok Kim *

Department of Mathematics, Korea University, Seoul 02841, Korea; jintae2002@korea.ac.kr (J.P.); there122@korea.ac.kr (S.Y.); chae1228@korea.ac.kr (C.L.)
* Correspondence: cfdkim@korea.ac.kr

Received: 16 May 2020; Accepted: 19 June 2020; Published: 22 June 2020

Abstract: In this article, we present a simple method for network visualization. The proposed method is based on distmesh [P.O. Persson and G. Strang, A simple mesh generator in MATLAB, SIAM Review 46 (2004) pp. 329–345], which is a simple unstructured triangular mesh generator for geometries represented by a signed distance function. We demonstrate a good performance of the proposed algorithm through several network visualization examples.

Keywords: Network; graph drawing; planar visualizations

1. Introduction

Since the formation of society, the relationships between its components have been significant. These relationships become more complex as society progresses; in addition, the components of society have also diversified. In sociology, a bundle of relationships is referred to as a network, which became a central concept in sociology in the 1970s. In a modern society called an information society, we have information regarding networks that has been transformed into concrete data. With a vast amount of information, information visualization has been used to analyze network and is gaining popularity. Techniques for information visualization have evolved, and they vary depending on the type of data [1–4]. Among the methods, visualization using graphs is one of the most helpful for understanding data and their relationships. The authors in [5] showed various graphs used in information visualization including tree layouts, H-tree layouts, balloon layout, radial layout, hyperbolic trees, fisheye graphs, and animated radial layouts (see Figure 1 as an example of network plot). Furthermore, toolkits for information visualization such as Prefuse, Protovis, and GUESS have been developed and widely used [1,6–8]. In several studies, nodes represent subjects, such as people and businesses, whereas edges represent relationships, such as friendships and partnerships. The scope of a network is not limited to people and institutions: if something is in an interactive relationship, we can call it a network, and networks can be also graphically identified by data. Network visualization is therefore being used in a variety of fields. For example, analysis for social and personal networks [9], pharmacological networks [10], biological networks [11,12], financial networks [13], and street networks [14] have been actively conducted.

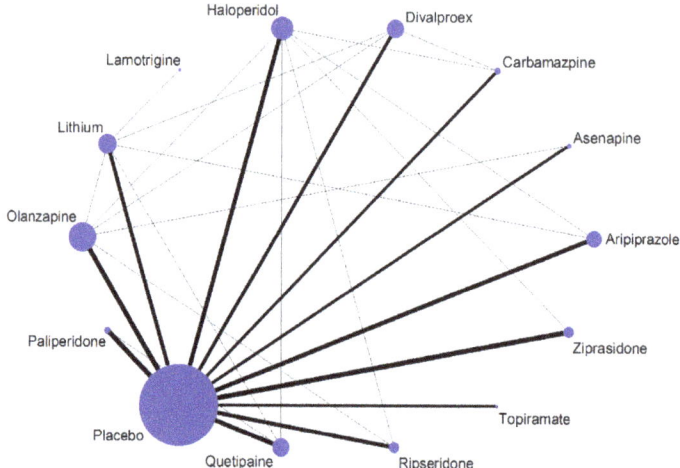

Figure 1. Example of a circular network. Reprinted from Salanti et al. [15] with permission from PLoS ONE.

Automatically drawing a network diagram requires algorithms. One of such algorithms is a classical force-directed algorithm that employs straight-edges. The force-directed algorithm treats edges as springs [16]. This algorithm turned the graph representation problem into a mathematical optimization problem. In other words, by reducing the energy generated by the spring system, we can find the equilibrium of the graph. The force-directed method has advantages such as simplicity to use and a good theoretical basis. As a result, many new methods of graph representation have been developed based on the method. As a typical example, Kamada and Kawai introduced an ideal distance for drawing graphs [17]. Let $\{X_1, X_2, \ldots, X_n\}$ be n-vertices and assume that they are spring-connected. The total energy of the spring is then expressed as follows:

$$\mathcal{E} = \sum_{i=1}^{n-1} \sum_{j=i+1}^{n} \frac{k_{ij}}{2}(|X_i - X_j| - l_{ij})^2,$$

where l_{ij} is the desirable length of the spring between X_i and X_j, k_{ij} is a parameter representing the strength of this spring, and $|\cdot|$ is the Euclidean norm. The desirable length represents the final length after executing the algorithm, and the strength of the spring refers to the tension of the spring keeping certain distance. The best graph is determined by minimizing $\mathcal{E}$. Please refer to [17] for more details about the algorithm and parameter definition. Another approach for automatically drawing a network diagram is based on the algorithm presented by Hall [18]. The main idea of this algorithm is to find the position of nodes $\{X_1, X_2, \ldots, X_N\}$ which minimizes

$$\mathcal{E} = \sum_{i<j}^{N} a_{ij}|X_i - X_j|^2, \qquad (1)$$

where $a_{ij} \geq 0$ is the connection weight between X_i and X_j. This algorithm is suitable for application to a structured data such as polyhedron [19]. However, it may not work well on actual data [20]. Rücker and Schwarzer et al. [20] introduced a method of automatically drawing network diagrams using graph theory and studied network meta-analysis. Furthermore, the algorithm was applied to a variety of examples from the literature. Another representative method for drawing network diagrams is the stress majorization [21]. The objective function is defined as follows:

$$\mathcal{E} = \sum_{i \neq j}^{N} w_{ij}(|\mathbf{X}_i - \mathbf{X}_j| - d_{ij})^2, \qquad (2)$$

where w_{ij} is the weight between $\mathbf{X}_i$ and $\mathbf{X}_j$, and d_{ij} is an ideal distance. For additional details about the algorithm, please refer to [21]. This algorithm was applied to real networks related to diseases and implemented by using the function *netgraph* in the R package *netmeta* [20].

We propose a simple algorithm for network visualization based on the distmesh algorithm [22] in this paper. The proposed method employs a distance d_{ij}, which is given by a reciprocal of weight w_{ij}, hence the computing process is essentially simple. Furthermore, the position of nodes is renewed proportionally by the net force, which is based on the gradient, therefore one can obtain an optimal diagram to the given data. A two-step stopping criterion is applied to further maximize the visual effect of the network diagram. Compared to other methods based on the gradient to optimize total level of movements, for instance, the force-directed method, the stress majorization method, etc., our proposed algorithm is simple to implement.

The contents of this article are organized as follows. In Section 2, the proposed algorithm is described. In Section 3, specific examples of network visualization are presented. Conclusions are presented in Section 4.

2. Numerical Algorithm

2.1. Distmesh Algorithm

A brief introduction to the distmesh algorithm [22], which is employed to generate the triangular mesh in domain with the level set representation, is presented in this section. We define the level set representation in the two-dimensional domain which imposes that the interface structure is treated as the zero-level set. The following procedure depicts the whole algorithm of the distmesh. A function $\psi(x,y) = \sqrt{x^2 + y^2} - 1$ is adopted to a sample level set description. Figure 2 depicts the overall process of distmesh algorithm quite in detail.

Step 1. Generate the random nodes $\mathbf{X}^0$ in domain.
Step 2. Generate a level set function ψ in the bounding box which includes the domain. The boundary of domain is regarded as the zero-level set.
Step 3. Perform the Delaunay triangulation with $\mathbf{X}^n$ if the maximal arrangement of nodes is greater than certain level. If $n = 0$, an initial Delaunay triangulation is accomplished. For the next step, compute the net force $\mathbf{F}$ in order to update the position of nodes.
Step 4. Renew the position of nodes to $\mathbf{X}^{n+1/2}$ by adding $\Delta t \mathbf{F}$.
Step 5. Push back the nodes that are pushed out to the boundary into the interface using the following equation

$$\mathbf{X}_i^{n+1} = \chi(\mathbf{X}_i^{n+1/2}) \left(\mathbf{X}_i^{n+1/2} - \frac{\nabla \psi(\mathbf{X}_i^{n+1/2})}{|\nabla \psi(\mathbf{X}_i^{n+1/2})|^2} \psi(\mathbf{X}_i^{n+1/2}) \right), \qquad (3)$$

where $\chi(\mathbf{X}_i^{n+1/2})$ is 1 if $\mathbf{X}_i^{n+1/2}$ is placed outside of the boundary; otherwise 0.
Step 6. Repeat **Step 3–5** until the level of the total movement of nodes is less than a given tolerance.

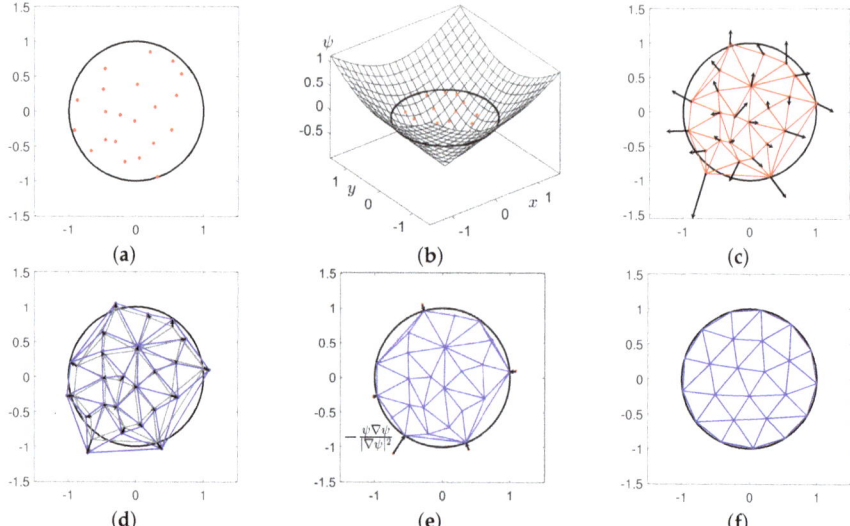

Figure 2. Schematic illustration of generating the distmesh. (**a**) Generated random nodes in the domain. (**b**) Signed distance function ψ in bounding box. The boundary of domain is regarded as the zero-level set. (**c**) Net force **F** in current triangulation. (**d**) Arrangement of nodes via $\Delta t \mathbf{F}$. (**e**) Projection of the nodes located outside $\psi > 0$ into the boundary $\psi \approx 0$ using Equation (3). (**f**) Final result of unstructured mesh by using the distmesh algorithm.

Using the distmesh algorithm, triangular mesh generation can be performed nonuniformly on domain of various shapes. The following Figure 3 is an example of such generated mesh.

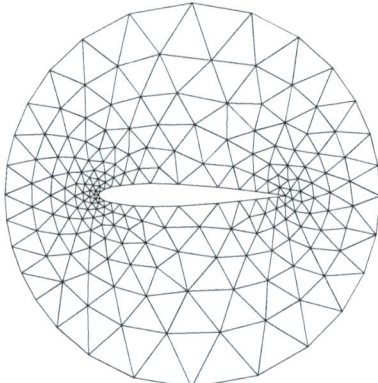

Figure 3. Example of nonuniformly generated mesh: the airfoil.

2.2. Proposed Algorithm for Network Visualization

The proposed algorithm for network visualization seeks to find $\{\mathbf{X}_1, \mathbf{X}_2, \ldots, \mathbf{X}_N\}$ that minimize the objective function

$$\mathcal{E} = \sum_{i<j}^{N} w_{ij} ||\mathbf{X}_i - \mathbf{X}_j| - d_{ij}|^2, \qquad (4)$$

where w_{ij} and d_{ij} are the weighting value and the desired distance between nodes X_i and X_j, respectively. The proposed algorithm is based on distmesh [22], which is a simple unstructured triangular mesh generator for geometries represented by a signed distance function. Let $\{X_1^n, X_2^n, \ldots, X_N^n\}$ be given node positions at iteration n. For simplicity of exposition, we assume $0 \leq w_{ij} \leq 1$. We then propose the following distance function:

$$d_{ij} = d(w_{ij}) = \frac{1}{w_{ij}^p}, \text{ for } w_{ij} > 0, \tag{5}$$

where p is a constant. Let minW be the minimum positive value of w_{ij}, i.e.,

$$\text{minW} = \min_{\substack{1 \leq i,j \leq N \\ w_{ij} > 0}} w_{ij}.$$

As shown in Figure 4, by setting the minimum distance minD = 1 when $w_{ij} = 1$ and the maximum distance maxD when $w_{ij} = \text{minW}$, we obtain

$$p = -\frac{\ln(\text{maxD})}{\ln(\text{minW})}. \tag{6}$$

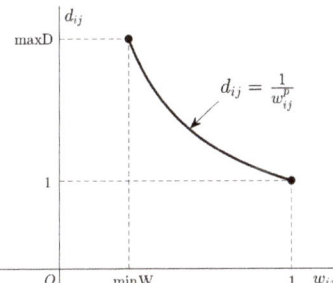

Figure 4. Illustration of distance function d_{ij} related to weighting value w_{ij}. minD = 1 and maxD are set to appear when $w_{ij} = 1$ and $w_{ij} = \text{minW}$, respectively.

Figure 5a,b show repulsive and attractive forces at nodes X_i^n and X_j^n when $|X_i^n - X_j^n| < d_{ij}$ and $|X_i^n - X_j^n| > d_{ij}$, respectively.

(a)   (b)

Figure 5. Two possible forces at nodes X_i^n and X_j^n: (a) repulsive force and (b) attractive force.

We loop over all the line segments connecting two nodes and compute the net force vector F_i^n at each node point X_i^n:

$$F_i^n = \sum_{\substack{j=1, j \neq i \\ w_{ij} > 0}}^{N} (|X_i^n - X_j^n| - d_{ij}) \frac{X_j^n - X_i^n}{|X_j^n - X_i^n|}.$$

Then, we update the position of the node points as

$$\mathbf{X}_i^{n+1} = \mathbf{X}_i^n + \Delta t \mathbf{F}_i^n, \quad \text{for } 1 \leq i \leq N, \qquad (7)$$

where Δt is an artificial time step. Upon updating the position of the node points, the network diagram is drawn automatically. The iterative algorithm has reached an equilibrium state if

$$\sqrt{\frac{1}{N} \sum_{i=1}^{N} |\mathbf{F}_i^k|^2} < tol_1 \qquad (8)$$

after k iterations.

As a concrete example, we consider three points $\mathbf{X}_1$, $\mathbf{X}_2$, and $\mathbf{X}_3$. Assume that the weighting matrix between $\mathbf{X}_i$ and $\mathbf{X}_j$ is given as

$$\mathbf{W} = \begin{pmatrix} 0 & 2 & 4 \\ 2 & 0 & 1 \\ 4 & 1 & 0 \end{pmatrix}.$$

We scale the matrix $\mathbf{W}$ by dividing the elements by the maximum value among elements and redefine $\mathbf{W}$ as

$$\mathbf{W} = \begin{pmatrix} 0 & 0.5 & 1 \\ 0.5 & 0 & 0.25 \\ 1 & 0.25 & 0 \end{pmatrix}.$$

Let $\mathbf{X}_1^0 = (\frac{3}{4}, \frac{3\sqrt{3}}{4})$, $\mathbf{X}_2^0 = (0,0)$, and $\mathbf{X}_3^0 = (\frac{3}{2}, 0)$, where the superscript 0 denotes the starting index. Here, we use $\Delta t = 0.3$, minD= 1, maxD= 2, minW= 0.25, and $tol_1 = 0.01$. Consequently, we get $p = 0.5$ and

$$\begin{pmatrix} & d_{12} & d_{13} \\ d_{21} & & d_{23} \\ d_{31} & d_{32} & \end{pmatrix} = \begin{pmatrix} & \sqrt{2} & 1 \\ \sqrt{2} & & 2 \\ 1 & 2 & \end{pmatrix}.$$

Figure 6a indicates the position of the three points with red markers, and the non-zero elements of $\mathbf{W}$ are represented by gray lines. In particular, the values of each element is expressed by the thickness of the line. The red arrows are net force vectors $\mathbf{F}_1^0$, $\mathbf{F}_2^0$ and $\mathbf{F}_3^0$. Using these net force vectors, we update the positions as

$$\mathbf{X}_1^1 = \mathbf{X}_1^0 + \Delta t \mathbf{F}_1^0, \quad \mathbf{X}_2^1 = \mathbf{X}_2^0 + \Delta t \mathbf{F}_2^0, \quad \mathbf{X}_3^1 = \mathbf{X}_3^0 + \Delta t \mathbf{F}_3^0,$$

which are shown in Figure 6b. Figure 6c–e show the network diagrams after 2, 3, and 6 iterations, respectively. The equilibrium state of the network diagram is obtained after 10 iterations as shown in Figure 6f. Even though the nodes are initially arranged in an equilateral triangle with sides of length 1.5, the network diagram in equilibrium is drawn according to the given weights.

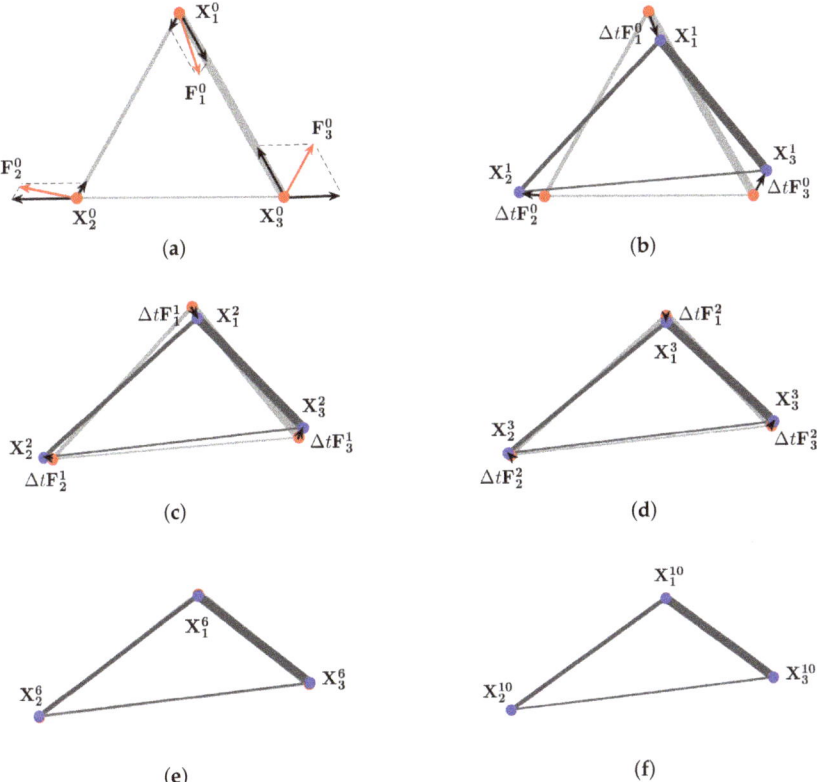

Figure 6. Schematic of the proposed algorithm. (**a**) initial condition, (**b**) after 1 iteration, (**c**) after 2 iterations, (**d**) after 3 iterations, (**e**) after 6 iterations, and (**f**) equilibrium state after 10 iterations.

3. Numerical Results

In this section, we present the generation of a network diagram with more data to confirm the efficiency and robustness of the proposed method. Specifically, we select 19 nodes and 19×19 matrix **W**, which are given in Appendix A. The matrix is created based on the dialogue between the characters in William Shakespeare's play, 'The Venice Merchant'. Each element w_{ij} of the matrix is the cumulative number of conversations between person i and person j. The parameters used are $\Delta t = 0.01$, minD $= 1$, maxD $= 2$, and $tol_1 = 0.01$. The value of p is then approximately 0.1879. Figure 7 shows process of the network visualization by our proposed method. The equilibrium state of the network diagram appears after 1985 iterations.

After 1985 iterations, each node is appropriately located according to the weights between the nodes in the network. This means that even if the nodes are initially randomly arranged, the network diagram is well drawn by our distance function. While the network plot is drawn, we can see that the objective function $\mathcal{E}$ is decreasing. As shown in Figure 8, $\mathcal{E}$ decreases and converges as time goes by. This shows that our proposed method has a mathematical basis for drawing the network diagram.

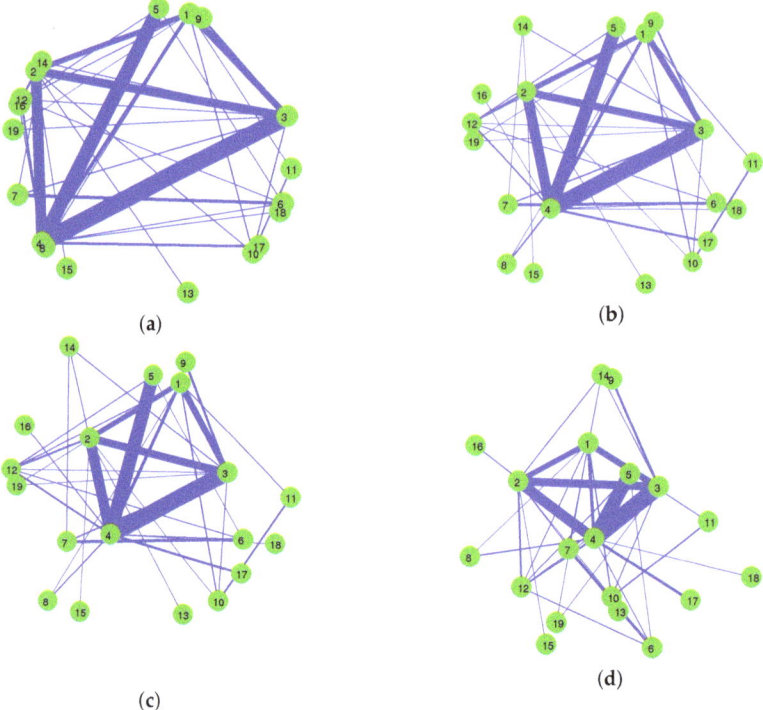

Figure 7. Snapshots of the network visualization process for 'The Venice Merchant': (**a**) initial condition, (**b**) after 20 iteration, (**c**) after 40 iterations, and (**d**) equilibrium state after 1985 iterations.

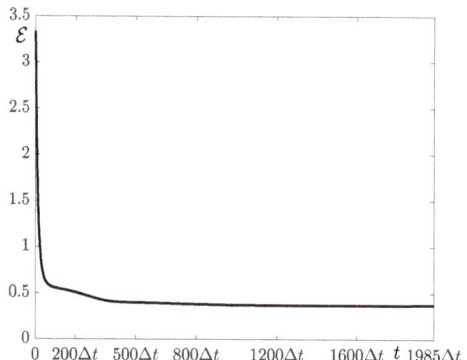

Figure 8. $\mathcal{E}$ decreases and converges while each node is moving.

However, the equilibrium state network diagram is not visually good. This is due to the nodes (9, 13, 15, 16, 17, 18) that have only one connection. Therefore, we further update the location of the nodes that have only one connection while fixing the other nodes. Let Ω_s be the index set of the nodes having only one connection. We compute the net force vector $\mathbf{F}_i^n$ at each node point $i \in \Omega_s$ as follows:

$$\mathbf{F}_i^n = \sum_{\substack{j=1,\ j\neq i \\ w_{ij}>0}}^{N} \frac{\mathbf{X}_i^n - \mathbf{X}_j^n}{|\mathbf{X}_i^n - \mathbf{X}_j^n|}.$$

Then, we temporally update the node points as

$$X_i^* = X_i^n + \Delta t F_i^n \text{ for } i \in \Omega_s, \tag{9}$$

where $\Delta t = 10$ is used. Finally, we set

$$X_i^{n+1} = X_j^n + d_{ij}\frac{X_i^* - X_j^n}{|X_i^* - X_j^n|} \text{ for } i \in \Omega_s \text{ and } X_i^{n+1} = X_i^n \text{ for } i \notin \Omega_s, \tag{10}$$

where X_j^n is the unique node connecting X_i^*. We define that the equilibrium state of the second step has been attained if

$$\sqrt{\frac{1}{|\Omega_s|}\sum_{i\in\Omega_s}|X_i^{k+1}-X_i^k|^2} < tol_2 \tag{11}$$

after k iterations, where $|\Omega_s|$ is the counting measure. Here, $tol_2 = 0.002$ is used. Therefore, the second step effectively rotates the node that has only one connection around the connecting node so that the overall distribution of the nodes is scattered.

Figure 9 illustrates the process of updating the position of nodes (red makers) that have only one connection. Figure 9a–d shows the network in the equilibrium state of the first step, after 1 iteration of the second step, after 2 iterations of the second step, and in the equilibrium state of the second step after 75 iterations.

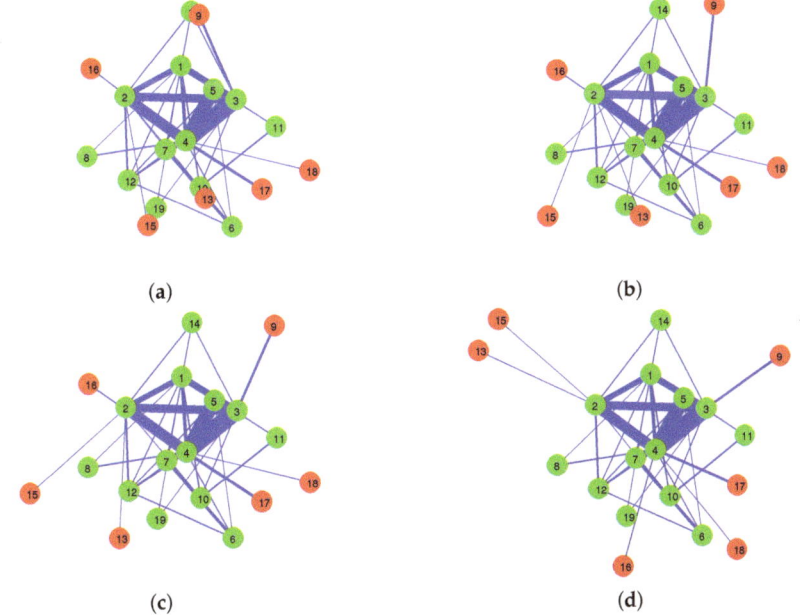

Figure 9. Updating the position of nodes with only one connection: (**a**) Equilibrium state of the first step, (**b**) after 1 iteration, (**c**) after 2 iterations, and (**d**) equilibrium state of the second step after 75 iterations.

Next, we consider another example 'Romeo and Juliet' which is a play written by William Shakespeare. Matrix **W** is defined by counting the number of conversations between 27 characters. The parameters used are minD = 1, maxD = 3, and $tol_1 = tol_2 = 0.002$, and then the value of p is approximately 0.2493. In particular, time step $\Delta t = 0.2$ and $\Delta t = 10$ are used in the first step and the second step, respectively. Figure 10a–c illustrate the character network at the initial condition, after the first step, and after the second step, respectively. From the results, we can find the main characters and relatively small parts.

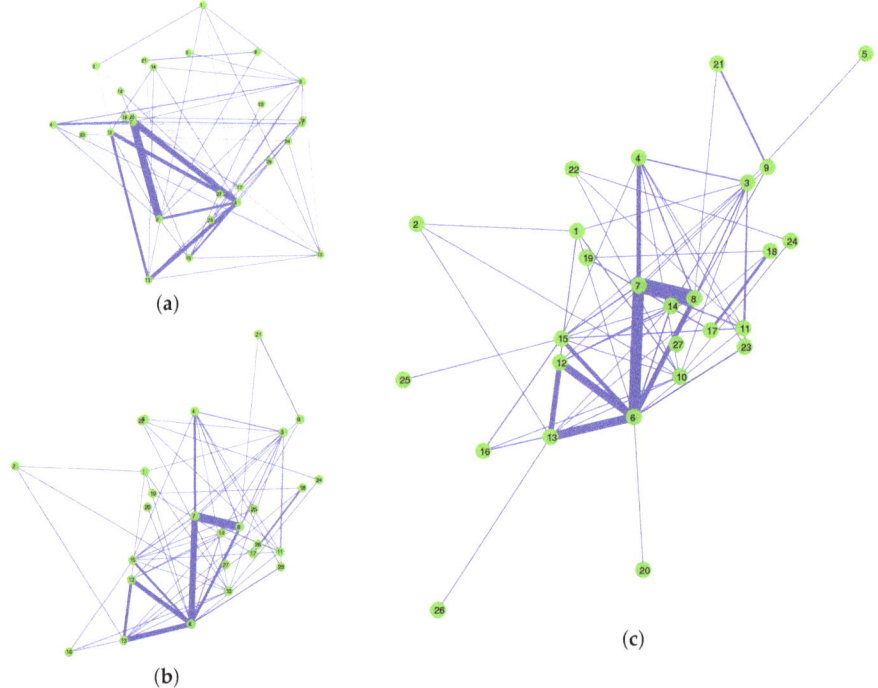

Figure 10. Snapshots of network visualization for 'Romeo and Juliet': (**a**) the initial condition, (**b**) after 230 iterations of the first step, and (**c**) after 20 iterations of the second step.

4. Conclusions

In this paper, we have proposed a simple method based on distmesh for network visualization. We have demonstrated the good performance of the proposed algorithm through network visualization examples. We can provide the MATLAB source code of this method for the interested readers. In future work, we plan to investigate effective network diagrams for character networks from novels and movies. We may further speed up the computation of the proposed method by using a Gauss–Newton–secant type method [23].

Author Contributions: All authors contributed equally to this work; J.P., S.Y., C.L. and J.K. critically reviewed the manuscript. All authors have read and agree to the published version of the manuscript.

Funding: The corresponding author (J. Kim) expresses thanks for the support from the BK21 PLUS program.

Acknowledgments: The authors thank the editor and the reviewers for their constructive and helpful comments on the revision of this article.

Conflicts of Interest: The authors declare no conflict of interest.

Appendix A

In this appendix, we provide the MATLAB source codes for network visualization. The following code is for 'The Merchant of Venice'. The code for 'Romeo and Juliet' is available on the following website:

http://elie.korea.ac.kr/~cfdkim/codes/

Listing A1: Matlab Code for the network visualization.

```matlab
% The first step
clear;
W=[ 0 21 24 16  0  0  0 2  0 7 4  5 0 0 0 0  0 0 1
   21  0 27 32  0  0  0 0  0 2 0 11 2 3 2 0  0 0 0
   24 27  0 40  0  0  7 0 12 6 0  2 0 5 0 0  0 0 2
   16 32 40  0 36  3  0 7  0 0 0 10 0 0 0 3 13 2 0
    0  0  0 36  0  2  0 0  0 0 0  4 0 0 0 0  0 0 0
    0  0  0  3  2  0 15 0  0 0 0  4 0 0 0 0  0 0 0
    0  0  7  0  0 15  0 0  0 0 0  0 3 0 0 0  0 0 0
    2  0  0  7  0  0  0 0  0 0 0  0 0 0 0 0  0 0 0
    0  0 12  0  0  0  0 0  0 0 0  0 0 0 0 0  0 0 0
    7  2  6  0  0  0  0 0  0 9 0  0 0 0 0 0  0 0 0
    4  0  0  0  0  0  0 0  9 0 0  0 0 0 0 0  0 0 0
    5 11  2 10  4  4  0 0  0 0 0  0 0 0 0 0  0 0 0
    0  2  0  0  0  0  0 0  0 0 0  0 0 0 0 0  0 0 0
    0  3  5  0  0  0  3 0  0 0 0  0 0 0 0 0  0 0 0
    0  2  0  0  0  0  0 0  0 0 0  0 0 0 0 0  0 0 0
    0  0  0  3  0  0  0 0  0 0 0  0 0 0 0 0  0 0 0
    0  0  0 13  0  0  0 0  0 0 0  0 0 0 0 0  0 0 0
    0  0  0  2  0  0  0 0  0 0 0  0 0 0 0 0  0 0 0
    1  0  2  0  0  0  0 0  0 0 0  0 0 0 0 0  0 0 0];
N=size(W,1); rand("seed",3773); t=rand(N,1);
xy=[cos(2*pi*t),sin(2*pi*t)]; W=W/max(max(W));
minW=min(min(W(W>0))); minD=1; maxD=2; p=-log(maxD)/log(minW);
for i=1:N
for j=1:N
if W(i,j)>0
d(i,j)=1/W(i,j)^p;
end
end
end
dt=0.01; tol=0.01; n=0; error=2*tol;
while error≥tol
n=n+1; F = zeros(N,2);
for i=1:N
for j=i+1:N
if W(i,j)>0
vt = xy(j,:)-xy(i,:);
F(i,:) = F(i,:) + (norm(vt)-d(i,j))*vt/norm(vt);
F(j,:) = F(j,:) - (norm(vt)-d(i,j))*vt/norm(vt);
end
end
end
xy = xy + dt*F; error=norm(F)/sqrt(N);
if n==1 || mod(n,10)==0 || error<tol
figure(1); DrawNetwork(xy,W); pause(0.1)
end
end

% The second step
z=find(sum(W>0)==1); M=length(z);
for k=1:M
s(k)=find(W(z(k),:)>0);
```

```
end
xy0=xy; n=0; dt=10.0; tol=0.002; error=2*tol;
while error≥tol
n=n+1; F = zeros(N,2);
for k=1:M
v=[0 0];
for j=1:N
vt = xy(z(k),:)-xy(j,:);
if norm(vt)>0
v=v+vt/norm(vt);
end
end
F(z(k),:)=v/norm(v);
end
xy = xy + dt*F;
error=0;
for k=1:M
v=xy(z(k),:)-xy(s(k),:);
xy(z(k),:)=xy(s(k),:)+d(z(k),s(k))*v/norm(v);
error=error+norm(xy(z(k),:)-xy0(z(k),:))^2;
end
error=sqrt(error/M); xy0=xy;
figure(2); DrawNetwork(xy,W); pause(0.1)
end
```

Listing A2: Function code for DrawNetwork.

```
function DrawNetwork(xy,W)
N=length(xy); clf; hold on
for i=1:N
for j=i+1:N
if W(i,j)>0
plot(xy([i,j],1),xy([i,j],2),"b","linewidth",15*W(i,j)^2+1);
end
end
end
scatter(xy(:,1),xy(:,2),400,"g","filled");
for i = 1:N
text(xy(i,1)-0.04,xy(i,2),num2str(i));
end
axis off; axis image;
end
```

References

1. Heer, J.; Card, S.K.; Landay, J.A. Prefuse: A toolkit for interactive information visualization. In Proceedings of the SIGCHI Conference on Human Factors in Computing Systems, Portland, OR, USA, 2–7 April 2005; pp. 421–430.
2. Keim, D.A. Information visualization and visual data mining. *IEEE Trans. Vis. Comput. Graph.* **2002**, *8*, 1–8. [CrossRef]
3. McGuffin, M.J. Simple algorithms for network visualization: A tutorial. *Tsinghua Sci. Technol.* **2012**, *17*, 383–398. [CrossRef]

4. Van Wijk, J.J.; Van de Wetering, H. Cushion treemaps: Visualization of hierarchical information. In Proceedings of the 1999 IEEE Symposium on Information Visualization (InfoVis' 99), San Francisco, CA, USA, 24–29 October 1999; pp. 73–78.
5. Herman, I.; Melançon, G.; Marshall, M.S. Graph visualization and navigation in information visualization: A survey. *IEEE Trans. Vis. Comput. Graph.* **2000**, *6*, 24–43. [CrossRef]
6. Adar, E. GUESS: A language and interface for graph exploration. In Proceedings of the SIGCHI Conference on Human Factors in Computing Systems, Montréal, QC, Cananda, 22–27 April 2006; pp. 791–800.
7. Bostock, M.; Heer, J. Protovis: A graphical toolkit for visualization. *IEEE Trans. Vis. Comput. Graph.* **2009**, *15*, 1121–1128. [CrossRef] [PubMed]
8. Wylie, B.; Baumes, J. A unified toolkit for information and scientific visualization. In *Visualization and Data Analysis 2009*; SPIE: San Jose, CA, USA, 2009; p. 72430H.
9. McCarty, C.; Molina, J.L.; Aguilar, C.; Rota, L. A comparison of social network mapping and personal network visualization. *Field Methods* **2007**, *19*, 145–162. [CrossRef]
10. Nüesch, E.; Häuser, W.; Bernardy, K.; Barth, J.; Jüni, P. Comparative efficacy of pharmacological and non-pharmacological interventions in fibromyalgia syndrome: Network meta-analysis. *Ann. Rheum. Dis.* **2013**, *72*, 955–962. [CrossRef] [PubMed]
11. Wu, L.; Li, M.; Wang, J.X.; Wu, F.X. Controllability and Its Applications to Biological Networks. *J. Comput. Sci. Technol.* **2019**, *34*, 16–34. [CrossRef]
12. Xia, M.; Wang, J.; He, Y. BrainNet Viewer: A network visualization tool for human brain connectomics. *PLoS ONE* **2013**, *8*, e68910. [CrossRef] [PubMed]
13. Dolfin, M.; Knopoff, D.; Limosani, M.; Xibilia, M.G. Credit Risk Contagion and Systemic Risk on Networks. *Mathematics* **2019**, *7*, 713. [CrossRef]
14. Pueyo, O.; Pueyo, X.; Patow, G. An overview of generalization techniques for street networks. *Graph. Models* **2019**, *106*, 101049. [CrossRef]
15. Chaimani, A.; Higgins, J.P.; Mavridis, D.; Spyridonos, P.; Salanti, G. Graphical tools for network meta-analysis in STATA. *PLoS ONE* **2013**, *8*, e76654. [CrossRef] [PubMed]
16. Eades, P. A heuristic for graph drawing. *Congr. Numer.* **1984**, *42*, 149–160.
17. Kamada, T.; Kawai, S. An algorithm for drawing general undirected graphs. *Inf. Process. Lett.* **1989**, *31*, 7–15. [CrossRef]
18. Hall, K.M. An *r*-dimensional quadratic placement algorithm. *Manag. Sci.* **1970**, *17*, 219–229. [CrossRef]
19. Spielman, D. Spectral Graph Theory. In *Combinatorial Scientific Computing (No. 18)*; CRC Press: Boca Raton, FL, USA, 2012.
20. Rücker, G.; Schwarzer, G. Automated drawing of network plots in network meta-analysis. *Res. Synth. Methods* **2016**, *7*, 94–107. [CrossRef] [PubMed]
21. Gansner, E.R.; Koren, Y.; North, S. Graph drawing by stress majorization. In *International Symposium on Graph Drawing*; Springer: Berlin/Heidelberg, Germany, 2004; pp. 239–250.
22. Persson, P.O.; Strang, G. A simple mesh generator in MATLAB. *SIAM Rev. Soc. Ind. Appl. Math.* **2004**, *46*, 329–345. [CrossRef]
23. Argyros, I.; Shakhno, S.; Shunkin, Y. Improved Convergence Analysis of Gauss-Newton-Secant Method for Solving Nonlinear Least Squares Problems. *Mathematics* **2019**, *7*, 99. [CrossRef]

© 2020 by the authors. Licensee MDPI, Basel, Switzerland. This article is an open access article distributed under the terms and conditions of the Creative Commons Attribution (CC BY) license (http://creativecommons.org/licenses/by/4.0/).

Article

Numerical Solution of the Cauchy-Type Singular Integral Equation with a Highly Oscillatory Kernel Function

SAIRA [1,†], Shuhuang Xiang [1,*,†] and Guidong Liu [2]

[1] School of Mathematics and Statistics, Central South University, Changsha 410083, China; sairahameed@csu.edu.cn
[2] School of Statistics and Mathematics, Nanjing Audit University, Nanjing 211815, China; liugd@nau.edu.cn
* Correspondence: xiangsh@csu.edu.cn; Tel.: +86-1397-314-3907
† Current address: School of Mathematics and Statistics, Central South University, Changsha 410083, China.

Received: 16 August 2019; Accepted: 18 September 2019; Published: 20 September 2019

Abstract: This paper aims to present a Clenshaw–Curtis–Filon quadrature to approximate the solution of various cases of Cauchy-type singular integral equations (CSIEs) of the second kind with a highly oscillatory kernel function. We adduce that the zero case oscillation ($k = 0$) proposed method gives more accurate results than the scheme introduced in Dezhbord et al. (2016) and Eshkuvatov et al. (2009) for small values of N. Finally, this paper illustrates some error analyses and numerical results for CSIEs.

Keywords: Clenshaw–Curtis–Filon; high oscillation; singular integral equations; boundary singularities

1. Introduction

Integral equations have broad roots in branches of science and engineering [1–6]. Cauchy-type singular integral equations (CSIEs) of the second kind occur in electromagnetic scattering and quantum mechanics [7] and are defined as:

$$au(x) + \frac{b}{\pi} \fint_{-1}^{1} \frac{u(y)K(x,y)}{y-x} dy = f(x), \quad x \in (-1,1). \tag{1}$$

A singular integral equation with a Cauchy principal value is a generalized form of an airfoil equation [8]. Here a and b are constants such that $a^2 + b^2 = 1$, $b \neq 0$ and $K(x,y) = e^{ik(y-x)}$ are the highly oscillatory kernel function. The function $f(x)$ is the Hölder continuous function, whereas $u(x)$ is an unknown function. The solution to the above-mentioned Equation (1) contains boundary singularities $w(x) = (x+1)^\alpha(1-x)^\beta$, i.e., $u(x) = w(x)g(x)$ and $g(x)$ is a smooth function [9,10]. Then the above Equation (1) transforms into:

$$aw(x)g(x) + \frac{b}{\pi} \fint_{-1}^{1} \frac{w(y)g(y)e^{ik(y-x)}}{y-x} dy = f(x), \quad x \in (-1,1), \tag{2}$$

where $\alpha, \beta \in (-1,1)$ depend on a and b, such that:

$$\alpha = \frac{1}{2\pi i} \log\left(\frac{a-ib}{a+ib}\right) - N, \quad \beta = \frac{-1}{2\pi i} \log\left(\frac{a-ib}{a+ib}\right) - M, \tag{3}$$

$$\kappa = -(\alpha + \beta) = M + N.$$

Here M and N are integers in $[-1,1]$, whereas the index of the integral equation is called κ, analogous to a class of functions, wherein the solution is to be sought. It is pertinent to mention that to produce integrable singularities in the solution, the index κ is restricted to three cases, $[-1, 0, 1]$, but the addressed paper considers only two cases for κ, i.e., $\kappa \leq 0$. The value of the index κ depends on different values for M and N [11–13]. A great number of real life practical problems, e.g., for $\kappa = -1$, the so-called natched half-plane problem and another problem of a crack parallel to the free boundary of an isotropic semi-infinite plane, that can be reduced to Cauchy singular integral equations are addressed in [14–17]. Writing Equation (2) in operator form, we get [18]:

$$Hg = f, \qquad (4)$$

where:

$$Hg = aw(x)g(x) + \frac{b}{\pi} \int_{-1}^{1} \frac{w(y)g(y)e^{ik(y-x)}}{y-x} dy.$$

Let us define another operator:

$$H'f = aw^*(x)f(x) - \frac{b}{\pi} \int_{-1}^{1} \frac{w^*(y)f(y)e^{ik(y-x)}}{y-x} dy, \qquad (5)$$

further:

$$\begin{aligned} HH' &= I \quad \text{if } \kappa > 0 \\ HH' &= H'H = I \quad \text{if } \kappa = 0 \\ H'H &= I \quad \text{if } \kappa < 0 \end{aligned} \qquad (6)$$

where $w^*(x) = (1+x)^{-\alpha}(1-x)^{-\beta}$.

It is worthy mentioning the fact that the solution for CSIE exists but unfortunately it is not unique, as CSIE has three solution cases for different values of κ. The aforementioned theorem appertains to the existence of the solution of CSIE for case $\kappa = 0$.

Theorem 1. *[13,15] (Existence of CSIEs) Let the singular integral Equation (2) be equivalent to a Fredholm integral equation, which implies that every solution of a Fredholm integral equation is the solution of a singular integral equation and vice versa.*

Proof. Based on Equations (4)–(6) the SIE (2) can be transforms into:

$$g = H'f.$$

Furthermore, it can be written as a Fredholm integral equation:

$$u(y) + \int_{-1}^{1} N(y, \tau)y(\tau)d\tau = F(y). \qquad (7)$$

where:

$$F(y) = \frac{b}{\pi} w(y) \int_{-1}^{1} \frac{w^*(x)f(x)}{y-x} dx,$$

and:

$$N(y, \tau) = aK(x, \tau)w^{-1} - \frac{b}{\pi} w(y) \int_{-1}^{1} \frac{w^*(x)K(x, \tau)}{y-x} dx.$$

Thus the claimed theorem is proven. □

Moreover, for Equation (1) we have three cases for κ:

$$\kappa = \begin{cases} 1, & \alpha < 0, -1 < \beta, \quad \alpha \neq \beta, \\ -1, & 0 < \beta, \alpha < 1, \quad \alpha \neq \beta, \\ 0, & \alpha = -\beta, \quad |\beta| \neq \frac{1}{2}. \end{cases} \quad (8)$$

Similarly, solution cases of the CSIE of the second type depending on values of κ are:

- 1: The solution $u(x)$ for $\kappa = 1$ is unbounded at both end points $x = \pm 1$:

$$u(x) = af(x) - \frac{bw(x)}{\pi} e^{-ikx} \int_{-1}^{1} \frac{w^*(y)f(y)e^{iky}}{y-x} dy + Cw(x), \quad (9)$$

where C is an arbitrary constant such that:

$$\int_{-1}^{1} u(y) e^{iky} dy = C. \quad (10)$$

Equation (2) gets infinitely many solutions but is unique for the above condition.

- 2: The solution $u(x)$ is bounded for $\kappa = 0$ at $x = \pm 1$ and unbounded at $x = \mp 1$:

$$u(x) = af(x) - \frac{bw(x)}{\pi} e^{-ikx} \int_{-1}^{1} \frac{w^*(y)f(y)e^{iky}}{y-x} dy, \quad (11)$$

Equation (2) gets a unique solution.

- 3: The solution $u(x)$ is bounded at both end points $x = \pm 1$ for $\kappa = -1$:

$$u(x) = af(x) - \frac{bw(x)}{\pi} e^{-ikx} \int_{-1}^{1} \frac{w^*(y)f(y)e^{iky}}{y-x} dy. \quad (12)$$

Equation (2) has no solution unless it satisfies the following condition:

$$\int_{-1}^{1} \frac{f(y)e^{iky}}{w(y)} dy = 0. \quad (13)$$

For many decades researchers have been struggling to find an efficient method to get these solutions. The Galerkin method, polynomial collocation method, Clenshaw–Curtis–Filon method and the steepest descent method are some of the eminent methods among many others for the solution of SIEs [19–24]. Moreover, Chakarbarti and Berge [25] for a linear function $f(x)$ gave an approximated method based on polynomial approximation and Chebyshev points. Z.K. Eshkuvatov [10] introduced the method taking Chebyshev polynomials of all four kinds for all four different solution cases of the CSIE. Reproducing the kernel Hilbert space (RKHS) method has been proposed by A. Dezhbord et al. [26]. The representation of solution u(x) is in the form of a series in reproducing kernel spaces.

This research work introduces the Clenshaw–Curtis–Filon quadrature to approximate the solution for various cases of a Cauchy singular integral equation of the second kind, Equation (1), at equally spaced points x_i. So the integral equation takes the form:

$$u_N(x_i) = af(x_i) - \frac{bw(x_i)}{\pi} e^{-ikx_i} \int_{-1}^{1} \frac{w^*(y)f_N(y)e^{iky}}{y-x_i} dy, \quad (14)$$

depending on the κ. Furthermore, the results of the numerical example are compared with [10,26] for $k = 0$. Comparison reveals that the addressed method gives a more accurate approximation than these methods, Section 4 provides this phenomena. The rest of the paper is organised as follows; Section 2 defines the numerical evaluation of the Cauchy integral in CSIE and approximates the solution at equally spaced points x_i. Section 3 represents some error analyses for CSIE. Section 4 concludes this paper by giving numerical results.

2. Description of the Method

The presented Clenshaw–Curtis–Filon quadrature to approximate the integral term $I(\alpha, \beta, k, x) = \int_{-1}^{1} \frac{w(y)f(y)e^{iky}}{y-x} dy$ consists of replacing function $f(y)$ by its interpolation polynomial $P_N(y)$ at Clenshaw–Curtis point set, $y_j = \cos\frac{j\pi}{N}, j = 0, 1, \cdots, N$. Rewriting the interpolation in terms of the Chebyshev series:

$$f(y) \approx P_N(y) = \sum_{n=0}^{N} {}''c_n T_n(y). \tag{15}$$

Here $T_n(y)$ is the Chebyshev polynomial of the first kind of degree n. Double prime denotes a summation, wherein the first and last terms are divided by 2. The FFT is used for proficient calculation of the coefficient c_n [27,28], defined as:

$$c_n = \frac{2}{N} \sum_{j=0}^{N} {}'' f(y_j) T_n(y_j).$$

Let it be that for any fixed x we can elect N s.t $x \notin \{y_j\}$; then the interpolation polynomial is rewritten in the form of a Chebyshev series as:

$$\tilde{P}_{N+1}(y) = \sum_{n=0}^{N+1} a_n T_n(y)$$

where a_n can be computed in $O(N)$ operations once c_n are calculated [27,29]. The Clenshaw–Curtis–Filon quadrature rule for integral $I(\alpha, \beta, k, x)$ is defined as:

$$I(\alpha, \beta, k, x) = \int_{-1}^{1} \frac{w(y)f(y)e^{iky}}{y-x} dy = \int_{-1}^{1} \frac{w(y)\tilde{P}_{N+1}(y)e^{iky}}{y-x} dy = \sum_{n=0}^{N+1} a_n M_n(\alpha, \beta, k, x), \tag{16}$$

where $M_n(\alpha, \beta, k, x) = \int_{-1}^{1} \frac{w(y)T_n(y)e^{iky}}{y-x} dy$ are the modified moments. The forthcoming subsection defines the method to compute the moments $M_n(\alpha, \beta, k, x)$ efficiently.

Computation of Moments

A well known property for $T_n(y)$ is defined as [30]:

$$\frac{T_n(y) - T_n(x)}{y-x} = 2 \sum_{j=0}^{n-1} {}' U_{n-1-j}(y) T_j(x) = 2 \sum_{j=0}^{n-1} {}' U_{n-1-j}(x) T_j(y), \tag{17}$$

where the prime indicates the summation whose first term is divided by 2 and $U_n(y)$ is the Chebyshev polynomial of the second kind.

$$\begin{aligned}
M_n(\alpha,\beta,k,x) &= \int_{-1}^{1} \frac{w(y)T_n(y)e^{iky}}{y-x}dy \\
&= \int_{-1}^{1} \frac{w(y)(T_n(y) - T_n(x) + T_n(x))e^{iky}}{y-x}dy \\
&= \int_{-1}^{1} \frac{w(y)(T_n(y) - T_n(x))e^{iky}}{y-x}dy + T_n(x)\int_{-1}^{1} \frac{w(y)e^{iky}}{y-x}dy \\
&= \int_{-1}^{1} w(y)\left(2\sum_{j=0}^{n-1}{}' U_{n-1-j}(x)T_j(y)\right)e^{iky}dy + T_n(x)\int_{-1}^{1} \frac{w(y)e^{iky}}{y-x}dy \\
&= 2\sum_{j=0}^{n-1}{}' U_{n-1-j}(x)\int_{-1}^{1} w(y)T_j(y)e^{iky}dy + T_n(x)\int_{-1}^{1} \frac{w(y)e^{iky}}{y-x}dy
\end{aligned} \quad (18)$$

Piessens and Branders [31] have addressed the fourth homogenous recurrence relation for the integral without singularity $\overline{M}_n(\alpha,\beta,k) = \int_{-1}^{1} w(y)T_j(y)e^{iky}dy$.

$$ik\overline{M}_{n+2} + 2(n+\alpha+\beta+2)\overline{M}_{n+1} - 2(2\alpha-2\beta+ik)\overline{M}_n - 2(n-\alpha-\beta-2)\overline{M}_{n-1} + ik\overline{M}_{n-2} = 0, \quad n \geq 2, \quad (19)$$

along with four initial values:

$$\begin{aligned}
\overline{M}_0^0 &= 2^{\alpha+\beta+1}e^{-ik}\frac{\Gamma(\alpha+1)\Gamma(\beta+1)}{\Gamma(\alpha+\beta+2)}F_1(\alpha+1;\alpha+\beta+2;2ik), \\
\overline{M}_1^0 &= M_0(x,\alpha+1,\beta,k) - M_0(x,\alpha,\beta,k), \\
\overline{M}_2^0 &= \frac{i}{k}[2(\alpha+\beta+2)M_1 - (2\alpha-2\beta+ik)M_0], \\
\overline{M}_3^0 &= \frac{i}{k}[2(\alpha+\beta+3)M_2 - (4\alpha-4\beta+ik)M_1 + 2(\alpha+\beta+1)M_0],
\end{aligned} \quad (20)$$

where $F_1(\alpha+1;\alpha+\beta+2;2ik)$ stands for confluent hypergeometric function of the first kind. Unfortunately the discussed recurrence relation for moments $\overline{M}_n(\alpha,\beta,k)$ is numerically unstable in the forward direction for $n > k$; in this sense by applying Oliver's algorithm these modified moments can be computed efficiently [31,32].

The integral $\int_{-1}^{1} \frac{w(y)e^{iky}}{y-x}dy$ is computed by the steepest descent method; the original idea was given by Huybrenchs and Vandewalle [33] for sufficiently high oscillatory integrals.

Proposition 1. *The Cauchy singular integral $\int_{-1}^{1} \frac{w(y)e^{iky}}{y-x}dy$ can be transformed into:*

$$\int_{-1}^{1} \frac{w(y)e^{iky}}{y-x}dy = S_{-1} - S_1 + i\pi w(x)e^{ikx} \quad (21)$$

where:

$$\begin{aligned}
S_{-1} &= i^{\alpha+1}e^{-ik}\int_0^{\infty} \frac{y^\alpha(2-iy)^\beta}{-1+iy-x}e^{-ky}dy \\
S_1 &= (-i)^{\beta+1}e^{ik}\int_0^{\infty} \frac{y^\beta(2+iy)^\alpha}{1+iy-x}e^{-ky}dy.
\end{aligned} \quad (22)$$

Proof. Readers are referred to [34] for more details. □

The generalized Gauss Laguerre quadrature rule can be used to evaluate the integrals S_{-1} and S_1 in the above equation by using the command lagpts in chebfun [35]. Let $\{y_j^\alpha, w_j^\alpha\}_{j=1}^k$ be the nodes and weights of the weight functions $y^\alpha e^{-y}$ and $\{y_j^\beta, w_j^\beta\}_{j=1}^k$ be the nodes and weights of the weight functions $y^\beta e^{-y}$ in accordance with the generalized Gauss Laguerre quadrature rule. Moreover, these integrals can be approximated by:

$$S_{-1} \approx Q_k = \left(\frac{i}{k}\right)^{\alpha+1} e^{-ik} \sum_{j=1}^k w_j^\alpha \frac{(2-(i/k)y_j^\alpha)^\beta}{(-1+(i/k)y_j^\alpha - x)}$$

$$S_1 \approx Q_k = \left(\frac{i}{k}\right)^{\beta+1} e^{ik} \sum_{j=1}^k w_j^\beta \frac{(2+(i/k)y_j^\beta)^\alpha}{(-1+(i/k)y_j^\beta - x)}. \tag{23}$$

$M_n(\alpha, \beta, k, x)$ is obtained by substituting Equations (19) and (21) into the last equality of Equation (18). Finally, together with Equations (16) and (14), the approximate solution:

$$u_N(x_i) = af(x_i) - \frac{bw(x_i)}{\pi} e^{-ikx_i} \sum_{n=0}^{N+1} a_n M_n(\alpha, \beta, k, x), \tag{24}$$

for CSIE (1) is derived for different solution cases at equally spaced points.

3. Error Analysis

Lemma 1. *[36,37] Let $f(x)$ be a Lipschitz continuous function on $[-1, 1]$ and $P_N[f]$ be the interpolation polynomial of $f(x)$ at $N + 1$ Clenshaw–Curtis points. Then it follows that:*

$$\lim_{N \to \infty} \|f - P_N[f]\|_\infty = 0. \tag{25}$$

In particular,

- *(i) if $f(x)$ is analytic with $|f(x)| \leq M$ in an ellipse ε_ρ (Bernstein ellipse) with foci ± 1 and major and minor semiaxis lengths summing to $\rho > 1$, then:*

$$\|f - P_N[f]\|_\infty \leq \frac{4M}{\rho^N(\rho - 1)}. \tag{26}$$

- *(ii) if $f(x)$ has an absolutely continuous $(\kappa_0 - 1)$st derivative and a κ_0th derivative $f^{(\kappa_0)}$ of bounded variation V_{κ_0} on $[-1,1]$ for some $\kappa_0 \geq 1$, then for $N \geq \kappa_0 + 1$:*

$$\|f - P_N[f]\|_\infty \leq \frac{4V_{\kappa_0}}{\kappa_0 \pi N(N-1)\cdots(N-\kappa_0+1)}. \tag{27}$$

Proposition 2. *[29] Suppose that $f(y) \in C^{R+2}[-1,1]$ with $R = \lceil \min\{\alpha, \beta\} \rceil$, then the error of the Clenshaw–Curtis–Filon quadrature rule for integral $I[f]$ satisfies:*

$$E_N = |I(\alpha, \beta, k, x) - I_N(\alpha, \beta, k, x)| = O(k^{-2-\min\{\alpha,\beta\}}), \quad k \to \infty. \tag{28}$$

Theorem 2. *Suppose that $u_N(x)$ is the approximate solution of $u(x)$ of CSIE for case $\kappa \leq 0$, then for error $|u(x) - u_N(x)|$, $x \in (-1, 1)$, the Clenshaw–Curtis–Filon quadrature is convergent, i.e.:*

$$\lim_{N \to \infty} |u(x) - u_N(x)| = 0. \tag{29}$$

Proof. Suppose that $x \notin Y_{N+1}$, $f \in C^2[-1,1]$ and let

$$Q(y) = \begin{cases} \frac{f(y)-f(x)}{y-x}, & y \neq x \\ f'(x), & y = x. \end{cases}$$

It is stated that $Q(y) \in C^1[-1,1]$ and $\|Q'\|_\infty \leq \frac{3}{2}\|f''\|_\infty$, in addition $R(y) = \frac{P_{N+1}(y)-f(x)}{y-x}$ is a polynomial of degree at most N. Then error for solutions $u(x)$ and $u_N(x)$ to CSIE for cases $\kappa \leq 0$ is defined as:

$$u(x) = af(x) - \frac{b}{\pi}e^{-ikx}w(x)\int_{-1}^{1}\frac{w^*(y)f(y)e^{iky}}{y-x}dy,$$

$$u_N(x) = af(x) - \frac{b}{\pi}e^{-ikx}w(x)\int_{-1}^{1}\frac{w^*(y)\bar{P}_{N+1}(y)e^{iky}}{y-x}dy.$$

Then:

$$|u(x) - u_N(x)| = \left|a(f(x)-f(x)) - \frac{b}{\pi}e^{-ikx}w(x)\int_{-1}^{1}w^*(y)(Q(y)-R(y))e^{iky}dy\right|$$

$$\leq \frac{b}{\pi}w(x)\int_{-1}^{1}w^*(y)dy\|Q(y)-R(y)\|_\infty$$

$$= D\|Q(y)-R(y)\|_\infty.$$

where $D = \frac{bw(x)2^{\alpha+\beta+1}\Gamma(\alpha+1)\Gamma(\beta+1)}{\pi\Gamma(\alpha+\beta+2)}$. □

4. Numerical Examples

Example 1. Let us consider the CSIE of the second kind:

$$\frac{u(x)}{\sqrt{2}} + \frac{1}{\sqrt{2}\pi}e^{-ikx}\int_{-1}^{1}\frac{u(y)e^{iky}}{y-x}dy = \frac{f(x)}{\sqrt{2}} \quad (30)$$

where $f(x) = \cos(x)$. For $x = 0.5$ and $a = b = \frac{1}{\sqrt{2}}$, we get values of $\alpha = 0.25$ and $\beta = 0.25$ from Equation (3) for $\kappa = 0$. The absolute error for $u(x)$ is presented in Tables 1 and 2 below.

Table 1. Absolute error for $\kappa = 0$, bounded at $x = 1$.

k	N = 5	N = 10	N = 20
50	4.6387×10^{-9}	3.9207×10^{-14}	1.1102×10^{-16}
100	1.0881×10^{-9}	4.9564×10^{-15}	0
1000	3.8093×10^{-11}	4.0030×10^{-16}	2.4825×10^{-16}
10,000	5.1593×10^{-13}	2.2204×10^{-16}	1.1102×10^{-16}

Table 2. Absolute error for $\kappa = 0$, bounded at $x = -1$.

k	N = 5	N = 10	N = 20
50	1.1156×10^{-9}	9.1854×10^{-15}	1.1102×10^{-16}
100	3.2791×10^{-10}	5.6610×10^{-16}	1.1102×10^{-16}
1000	1.7225×10^{-12}	2.2204×10^{-16}	2.2204×10^{-16}
10,000	7.3056×10^{-15}	3.3307×10^{-16}	3.3307×10^{-16}

Example 2. *The mixed boundary value problem is described in Figure 1.*

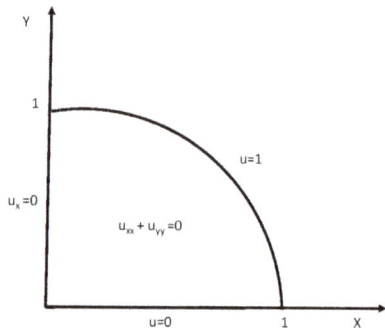

Figure 1. The mixed boundary value problem.

Taken from [18], it has the analytic solution $\phi(x,t) = \frac{2}{\pi}\arctan\frac{2y}{1-x^2-t^2}$. *It can further be reduced to the following integral equation for* $\kappa = -1$ *and for* $\alpha = \beta = \frac{1}{2}$.

$$\frac{-1}{\pi}\int_{-1}^{1}\frac{u(y)}{y-x}dy = C_1 + \frac{1}{\pi}\left[\frac{1-x}{2}\log(1-x) + \frac{1+x}{2}\log(1+x) - \log(2+x) - 1\right] \tag{31}$$

Here C_1 is a constant defined as $C_1 = 0.4192007182789807$. Furthermore if $u(x)$ is known, the solution of the above boundary value can be derived as:

$$\phi(\mu,\nu) = \frac{1}{\pi}\int_{-\infty}^{\infty}\frac{\nu u(y,0)}{(y-\mu)^2+\nu^2}dy$$

where:

$$u(y,0) = \begin{cases} u(y)+(1-y)/2, & |y| \le 1, \\ 1 & t \in [-2,-1], \\ 0, & otherwise. \end{cases} \tag{32}$$

So here we just solve $u(x)$ for simplicity. Figure 2 illustrates the absolute error for $u(x)$.

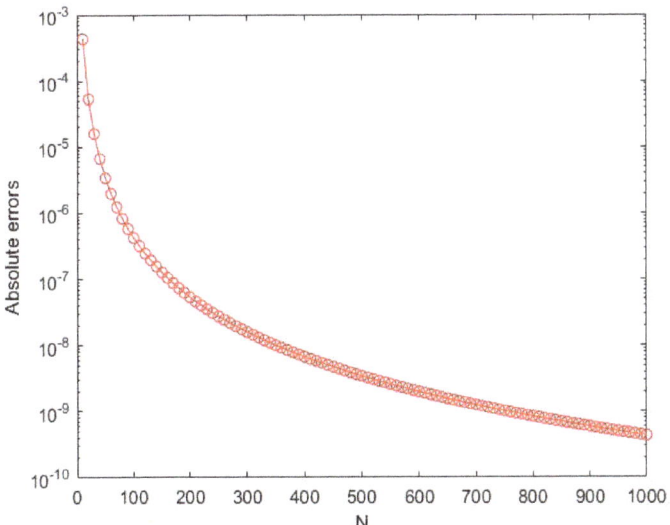

Figure 2. The absolute error for $u(x)$, for $x = 0.6$.

Figure 2 shows that absolute error for $u(x)$ decreases for greater values of N.

Example 3. *[10,26] For CSIE with $k = 0$:*

$$\int_{-1}^{1} \frac{u(y)}{y-x} dy = x^4 + 5x^3 + 2x^2 + x - \frac{11}{8} \tag{33}$$

in the case $a = 0$ and $b = 1$, where α and β are derived from Equation (3) and the exact values of $u(y)$ for cases $\kappa \leq 0$ for the solution bounded at $x = -1, x = 1, x = \pm 1$ are given as:

$$u(y) = \frac{1}{\pi} \sqrt{\frac{1+y}{1-y}} \left[y^4 + 4y^3 - 5/2y^2 + y - 7/2 \right]$$

$$u(y) = \frac{-1}{\pi} \sqrt{\frac{1-y}{1+y}} \left[y^4 + 6y^3 + 15/2y^2 + 6y + 7/2 \right] \tag{34}$$

$$u(y) = \frac{-1}{\pi} \sqrt{1-y^2} \left[y^3 + 5y^2 + 5/2y + 7/2 \right].$$

Table 3 presents the absolute error for the above three cases.

Table 3. Absolute error for case $\kappa \leq 0, k = 0$.

x	Error		
	$\kappa = -1$	$\kappa = 0$, boundned at $x = -1$	$\kappa = 0$, boundned at $x = 1$
−0.6	0	1.1102×10^{-16}	4.4409×10^{-16}
−0.2	3.3307×10^{-16}	2.2204×10^{-16}	4.4409×10^{-16}
0.2	2.2204×10^{-16}	4.4409×10^{-16}	0
0.6	0	2.2204×10^{-16}	4.4409×10^{-16}

Clearly, Table 3 shows that obtained absolute errors are significantly good for really small values of N, N = 5, that can never be achieved in [10,26]. The exact value for u(x) in the above examples is obtained through Mathematica 11, while the approximated results are calculated using Matlab R2018a on a 4 GHz personal laptop with 8 GB of RAM. For Example 2 Matlab code and Mathematica command is provided as supplementary material.

5. Conclusions

In the presented research work, the Clenshaw–Curtise–Filon quadrature is used to get higher order accuracy. Absolute errors are presented in Tables 1 and 2 for solutions of highly oscillatory CSIEs for $\kappa = 0$. For larger values of N, Figure 2 shows the absolute error for $u(x)$ for mixed the boundary value problem, whereas for frequency $k = 0$, the proposed quadrature posseses higher accuracy than the schemes claimed in [10,26]; Table 3 addresses this very well. This shows that the quadrature rule is quite accurate with the exact solution.

Supplementary Materials: The following are available online at http://www.mdpi.com/2227-7390/7/10/872/s1, for Example 2, Figure 2: The absolute error for $u(x)$, for $x = 0.6$.

Author Contributions: Conceptualization, SAIRA, S.X. and G.L.; Methodology, SAIRA; Supervision, S.X.; Writing (original draft), SAIRA; Writing (review and editing), SAIRA, S.X. and G.L.

Funding: This research received no external funding.

Conflicts of Interest: The authors declare no conflict of interest.

References

1. Polyanin, A.D.; Manzhirov, A.V. *Handbook of Integral Equations*; CRC Press: Boca Raton, FL, USA, 1998.
2. Li, J.; Wang, X.; Xiao, S.; Wang, T. A rapid solution of a kind of 1D Fredholm oscillatory integral equation. *J. Comput. Appl. Math.* **2012**, *236*, 2696–2705. [CrossRef]
3. Ursell, F. Integral equations with a rapidly oscillating kernel. *J. Lond. Math. Soc.* **1969**, *1*, 449–459. [CrossRef]
4. Yalcinbas, S.; Aynigul, M. Hermite series solutions of linear Fredholm integral equations. *Math. Comput. Appl.* **2011**, *16*, 497–506.
5. Fang, C.; He, G.; Xiang, S. Hermite-Type Collocation Methods to Solve Volterra Integral Equations with Highly Oscillatory Bessel Kernels. *Symmetry* **2019**, *11*, 168. [CrossRef]
6. Babolian, E.; Hajikandi, A.A. The approximate solution of a class of Fredholm integral equations with a weakly singular kernel. *J. Comput. Appl. Math.* **2011**, *235*, 1148–1159. [CrossRef]
7. Aimi, A.; Diligenti, M.; Monegato, G. Numerical integration schemes for the BEM solution of hypersingular integral equations. *Int. J. Numer. Method Eng.* **1999**, *45*, 1807–1830. [CrossRef]
8. Beyrami, H.; Lotfi, T.; Mahdiani, K. A new efficient method with error analysis for solving the second kind Fredholm integral equation with Cauchy kernel. *J. Comput. Appl. Math.* **2016**, *300*, 385–399. [CrossRef]
9. Setia, A. Numerical solution of various cases of Cauchy type singular integral equation. *Appl. Math. Comput.* **2014**, *230*, 200–207. [CrossRef]
10. Eshkuvatov, Z.K.; Long, N.N.; Abdulkawi, M. Approximate solution of singular integral equations of the first kind with Cauchy kernel. *Appl. Math. Lett.* **2009**, *22*, 651–657. [CrossRef]
11. Cuminato, J.A. Uniform convergence of a collocation method for the numerical solution of Cauchy-type singular integral equations: A generalization. *IMA J. Numer. Anal.* **1992**, *12*, 31–45. [CrossRef]
12. Cuminato, J.A. On the uniform convergence of a perturbed collocation method for a class of Cauchy integral equations. *Appl. Numer. Math.* **1995**, *16*, 439–455. [CrossRef]
13. Karczmarek, P.; Pylak, D.; Sheshko, M.A. Application of Jacobi polynomials to approximate solution of a singular integral equation with Cauchy kernel. *Appl. Math. Comput.* **2006**, *181*, 694–707. [CrossRef]
14. Lifanov, I.K. *Singular Integral Equations and Discrete Vortices*; Walter de Gruyter GmbH: Berlin, Germany, 1996.
15. Ladopoulos, E.G. *Singular Integral Equations: Linear and Non-Linear Theory and Its Applications in Science and Engineering*; Springer Science and Business Media: Berlin, Germany, 2013.
16. Muskhelishvili, N.I. *Some Basic Problems of the Mathematical Theory of Elasticity*; Springer Science and Business Media: Berlin, Germany, 2013.

17. Martin, P.A.; Rizzo, F.J. On boundary integral equations for crack problems. *Proc. R. Soc. Lond. A Math. Phys. Sci.* **1989**, *421*, 341–355. [CrossRef]
18. Cuminato, J.A. Numerical solution of Cauchy-type integral equations of index- 1 by collocation methods. *Adv. Comput. Math.* **1996**, *6*, 47–64. [CrossRef]
19. Asheim, A.; Huybrechs, D. Complex Gaussian quadrature for oscillatory integral transforms. *IMA J. Num. Anal.* **2013**, *33*, 1322–1341. [CrossRef]
20. Chen, R.; An, C. On evaluation of Bessel transforms with oscillatory and algebraic singular integrands. *J. Comput. Appl. Math.* **2014**, *264*, 71–81. [CrossRef]
21. Erdelyi, A. Asymptotic representations of Fourier integrals and the method of stationary phase. *SIAM* **1955**, *3*, 17–27. [CrossRef]
22. Olver, S. Numerical Approximation of Highly Oscillatory Integrals. Ph.D. Thesis, University of Cambridge, Cambridge, UK, 2008.
23. Milovanovic, G.V. Numerical calculation of integrals involving oscillatory and singular kernels and some applications of quadratures. *Comput. Math. Appl.* **1998**, *36*, 19–39. [CrossRef]
24. Dzhishkariani, A.V. The solution of singular integral equations by approximate projection methods. *USSR Comput. Math. Math. Phys.* **1979**, *19*, 61–74. [CrossRef]
25. Chakrabarti, A.; Berghe, G.V. Approximate solution of singular integral equations. *Appl. Math. Lett.* **2004**, *17*, 553–559. [CrossRef]
26. Dezhbord, A.; Lotfi, T.; Mahdiani, K. A new efficient method for cases of the singular integral equation of the first kind. *J. Comput. Appl. Math.* **2016**, *296*, 156–169. [CrossRef]
27. He, G.; Xiang, S. An improved algorithm for the evaluation of Cauchy principal value integrals of oscillatory functions and its application. *J. Comput. Appl. Math.* **2015**, *280*, 1–13. [CrossRef]
28. Trefethen, L.N. C hebyshev Polynomials and Series, Approximation theorey and approximation practice. *Soc. Ind. Appl. Math.* **2013**, *128*, 17-19.
29. Liu, G.; Xiang, S. Clenshaw Curtis type quadrature rule for hypersingular integrals with highly oscillatory kernels. *Appl. Math. Comput.* **2019**, *340*, 251–267. [CrossRef]
30. Wang, H.; Xiang, S. Uniform approximations to Cauchy principal value integrals of oscillatory functions. *Appl. Math. Comput.* **2009**, *215*, 1886–1894. [CrossRef]
31. Piessens, R.; Branders, M. On the computation of Fourier transforms of singular functions. *J. Comput. Appl. Math.* **1992**, *43*, 159–169. [CrossRef]
32. Oliver, J. The numerical solution of linear recurrence relations. *Numer. Math.* **1968**, *114*, 349–360. [CrossRef]
33. Huybrechs, D.; Vandewalle, S. On the evaluation of highly oscillatory integrals by analytic continuation. *Siam J. Numer. Anal.* **2006**, *44*, 1026–1048. [CrossRef]
34. Wang, H.; Xiang, S. On the evaluation of Cauchy principal value integrals of oscillatory functions. *J. Comput. Appl. Math.* **2010**, *234*, 95–100. [CrossRef]
35. Dominguez, V.; Graham, I.G.; Smyshlyaev, V.P. Stability and error estimates for Filon Clenshaw Curtis rules for highly oscillatory integrals. *IMA J. Numer. Anal.* **2011**, *31*, 1253–1280. [CrossRef]
36. Xiang, S.; Chen, X.; Wang, H. Error bounds for approximation in Chebyshev points. *Numer. Math.* **2010**, *116*, 463–491. [CrossRef]
37. Xiang, S. Approximation to Logarithmic-Cauchy Type Singular Integrals with Highly Oscillatory Kernels. *Symmetry* **2019**, *11*, 728.

© 2019 by the authors. Licensee MDPI, Basel, Switzerland. This article is an open access article distributed under the terms and conditions of the Creative Commons Attribution (CC BY) license (http://creativecommons.org/licenses/by/4.0/).

MDPI
St. Alban-Anlage 66
4052 Basel
Switzerland
Tel. +41 61 683 77 34
Fax +41 61 302 89 18
www.mdpi.com

Mathematics Editorial Office
E-mail: mathematics@mdpi.com
www.mdpi.com/journal/mathematics

www.ingramcontent.com/pod-product-compliance
Lightning Source LLC
LaVergne TN
LVHW070655100526
838202LV00013B/970